Theodor Billroth, J. Bentell Endean

The Care of the Sick at Home and in the Hospital

A Handbook for Families and for Nurses. Second Edition

Theodor Billroth, J. Bentell Endean

The Care of the Sick at Home and in the Hospital
A Handbook for Families and for Nurses. Second Edition

ISBN/EAN: 9783337022235

Printed in Europe, USA, Canada, Australia, Japan

Cover: Foto ©berggeist007 / pixelio.de

More available books at **www.hansebooks.com**

THE

CARE OF THE SICK

AT HOME AND IN THE HOSPITAL

A HANDBOOK

FOR FAMILIES AND FOR NURSES

BY

Dr TH. BILLROTH

PROFESSOR OF SURGERY IN VIENNA, ETC. ETC. ETC.

TRANSLATED BY SPECIAL AUTHORITY OF THE AUTHOR

BY

J. BENTALL ENDEAN

"Shall I succeed?
That is an idle question! It benefits mankind.
Instead of asking,
Venture!"

WITH PORTRAIT AND 51 ILLUSTRATIONS

SECOND EDITION

LONDON
SAMPSON LOW, MARSTON & Co., Ltd.
St. Dunstan's House
FETTER LANE, FLEET STREET, E.C.

1 8 9 1

To the Memory

OF

MY BELOVED MOTHER,

REVERENTLY AND LOVINGLY

I Dedicate this Volume.

J. BENTALL ENDEAN.

" One touch of nature makes the whole world kin."

" Troilus and Cressida."

"But mankind—the race would perish did they cease to aid each other. From the time that the mother binds the child's head till the moment that some kind assistant wipes the death-damp from the brow of the dying, we cannot exist without mutual help. All, therefore, that need aid, have right to ask it of their fellow-mortals; no one who has the power of granting can refuse it without guilt."—SIR WALTER SCOTT.

CONTENTS.

CHAPTER I.

CHAPTER II.

CHAPTER III.

CHAPTER IV.

CHAPTER V.

CONTENTS.

ILLUSTRATIONS.

PREFACE.

ON Sick-Nursing many books have been published, but I can recommend only the following :—

Dr L. G. Courvoisier. Die häusliche Krankenpflege. Basel, 1876.

Florence S. Lees. Handbuch für Krankenpflegerinnen, übersetzt von Dr Paul Schliep. Berlin, 1874.

Florence Nightingale. Rathgeber für Gesundheits- und Krankenpflege, übersetzt von Dr Paul Niemeyer. Leipzig, 1878.

Taschenbuch für Krankenpflegerinnen. Weimar, Jahrgänge 1879, 1880, und 1881.

Dr J. Wiel. Diätetisches Kochbuch ; in vielen Auflagen in Freiburg (Fr. Wagner'sche Buchhandlung) erschienen.

Dr J. Wiel. Tisch für Magenkranke ; ebenfalls in vielen Auflagen in Carlsbad (bei H. Feller) erschienen.

These books and others like them, although ably written, contain, in my opinion, either too little or too much. I consider that all systematic, anatomical, and physiological expositions in such books are just as superfluous as the systematic explanations of the ætiology and the diagnosis of every possible disease would be. That portion of medical science and art which can be briefly presented in a popular form, must be extremely superficial and incomplete. For an able, aspiring woman it is too little ; for a woman who seeks and finds her calling and happiness in sick-nursing, it is superfluous. To speak of patients and the care of them without saying anything of the *most important* diseases, their causes, and phenomena, is not sufficient; in this book, therefore, so much upon these subjects will be given as is indispensable to their full comprehension.

The nurse must be the helper of the patient and the doctor; she must learn to execute judiciously and accurately the doctor's orders, and must not desire to cure on her own account; she must have as implicit confidence in him as the patient himself has; when this is not the case, she will often be tempted to criticise his instructions by her own superficial knowledge, and, in her opinion, even to improve upon them. Such a nurse not only renders the practice of his profession more difficult to the doctor, but, under such circumstances, the patient himself always suffers the most; his confidence in his medical attendant having been shaken by the nurse, he tries first one treatment and then another, but none fully and regularly, until at last it may even happen that his life is sacrificed.

In the German method of treatment of the sick, particularly in hospitals, one great advantage is, that the head doctor and his assistants themselves do as much as possible—in wound-dressing this is of material consequence. Infinite care is necessary in dressing (bandaging) the wounded, and patients who have been operated on; every young doctor is not qualified for it, and some may *never* be qualified, because they are incapable of closely concentrating their attention upon technical manipulation. In the practice of the modern principles of dressing, the personal responsibility of the dresser has become an anxious responsibility. In England, France, Italy, Russia the doctors leave wound-dressing almost exclusively to the nurses.

All sorts of information on various methods of dressing (bandaging) given in other handbooks for nurses, will, in this book, be sought for in vain. Far from desiring to assert that, properly instructed, a woman is not able to execute a good dressing, however complicated, as carefully as a young doctor, yet, from experience, I know innumerable cases of women who possessed all the qualifications of an excellent nurse, but who were not quite capable of all that a doctor must learn and know.

Explanations that might wound feminine delicacy at the very beginning, I have avoided. Much is to be learnt quickly by the sick-bed, and many things will be easily overcome by degrees, the details of which, if printed, would have a deterrent effect.

It will, perhaps, be asked, after having omitted such subjects as copiously fill many chapters in other books for female nurses, what then remains? I hope my lady-readers will yet find in this book enough to learn and to impress permanently upon their memories.

For those who desire to qualify as Professional Sick-Nurses the contents of this Handbook, in my opinion, should constitute the First Lectures (of the Preparatory Course) on the Care of the Sick. It is evident that such a Preparatory Course must be combined simultaneously with practical demonstrations; it is equally a matter of course that the Trained Nurse must learn, by degrees, more than this book contains. How much the doctor will entrust to individual, specially-qualified, women, will depend upon himself. In the same way he must judge of the suitability of a woman for the post of Head-Nurse in a Hospital and Teacher of the Probationers; length of service must never decide it, but special qualification alone. For this post, certain qualities of character and of individuality are more necessary than an unusual degree of medical knowledge. Practical intelligence, clear comprehension of circumstances, knowledge of mankind, impartiality, patience, benevolence, quiet dignity, combined with firmness of character, make the woman, as the man, better fitted to be the leader of others, than great educational attainments do. These qualities are not obtained from books—they are partly innate, and partly to be acquired by long years of experience and self-training.

Let every woman and every girl interested in the Care of the Sick cheerfully take this book into their hands. In it they will find, I hope, many things that will enable them to help and benefit others.

But one charge will be brought against me: that I have written much that goes beyond the ordinary thinking powers of nurses in general. This charge I willingly accept. A similar charge was made against me when I wrote a book for students, and yet I have every reason to be satisfied with the success of that book. Whilst not writing for Professional Nurses only, but also for Educated Housewives and Mothers of Families who desire to have, not only something to learn, but also something to think about, and, apart from the consideration that women of small intellectual capacity are not generally fitted for nurses, I have always found that practical results in tuition are better gained by a sure method and gradual development of a well-regulated mind than by descending to the level of the deepest ignorance and the most inferior endowments. Women of mediocre intellect should rather be dissuaded from the nurse's calling, not simply because experience proves that the *reliable* execution of important duties absolutely presupposes a certain education of the intellect, but because the greatest kindness of heart can obtain practical beneficial results in the department of Sick-Nursing only when it is united with developed intelligence.

DR TH. BILLROTH.

VIENNA.

PREFACE TO THE THIRD EDITION.

Whilst the Second Edition of this book was a simple reprint of the First, it appeared desirable that some alteration should be made in this edition, the editing of which the author entrusted to me.

I would only refer to two additions by which the book has been enriched : one, the Appendix—" On the Structure and Functions of the Human Body," by Dr Billroth ; previously, this was published separately, and was given to those who attended the " Course on Nursing," held yearly in the Rudolfiner Haus, Vienna ; the other, the Illustrations of Bandaging as practised at these Courses.

It was desirable to revise Chapters V. and VI. ; this duty was kindly undertaken by Dr Karl Bettelheim, Senior Physician at the Rudolfiner Haus.

Dr R. GERSUNY.

Vienna, *September* 1888.

TRANSLATOR'S PREFACE.

Co-existent with life is the possibility of pain. The intuitive consciousness of this possibility is the chief safeguard of the higher organisms. Pain is the active expression of disease, whether that disease be induced within the organism itself, or whether it arise from external causes.

Man desires immunity from suffering—in all ages he has striven to secure it, and much knowledge conducive to this end has been obtained. This century marks an eventful era in the history of therapeutics, and the Surgeon and the Physician are now able to achieve vastly more in preventing and in lessening pain than was ever before possible—the results are universally beneficial. Everywhere the human organism is the same, is subject to like diseases and infirmities, and hence that which alleviates or overcomes these in one country is adapted to alleviate or overcome them in all.

From Dr Billroth's great experience and knowledge of the different methods employed in nursing the sick, he, like many other members of the medical profession, realized how little, comparatively, is known of the essential requirements of the sick-room; he therefore wrote this book for the instruction and guidance of all interested in the care of the sick, stating therein with much clearness, the causes, nature, and symptoms of various diseases, and the main principles of good sick-nursing. From these instructions a mother will learn, not only what is necessary to the care of a sick member of her family, but also the best means to adopt for the prevention of sickness and for the maintenance of health in the household.

It is the skilled nurse who, with sure but gentle hand, changes the dressing of the sensitive patient without causing unnecessary pain : her very touch is soothing to him ; she shakes his pillows and places him in a comfortable position, so that he feels rested and refreshed after her ministrations; even her voice is music as she cheers him, whilst going about her duties with a silent step ; and her presence in the sick-room is like a ray of sunlight, the messenger of hope and health.

In many homes the mission of the intelligent nurse is two-fold : first, to care for the sick ; second, to teach the close affinity that exists between cleanliness and health, dirt and disease.

The Portrait of Dr Billroth, by J. Löwy, K. u K. Hof-Photograph, Vienna, the Illustration of "The Billroth Bed-Crane," and the Index, &c., have been specially prepared for this Edition.

With the kind permission of Miss Florence Nightingale, the Extracts from "Notes on Nursing" are given as quoted by Dr Billroth.

I desire to express my best thanks to Fr. Goll, M.D., Professor of Pharmacology at the University of Zürich ; to Dr R. Gersuny, of Vienna ; to W. Hills, Esq. (Messrs J. Bell & Co., Oxford Street) ; and to other very highly esteemed friends for their valuable assistance in preparing this Edition.

J. BENTALL ENDEAN.

CANTERBURY,
September 21, 1890.

" Obedience is the sacrifice of Angels."
                                    SHAKESPEARE.

" He holds no parley with unmanly fears ;
   Where duty bids, he confidently steers,
   Faces a thousand dangers at her call,
   And, trusting in his God, surmounts them all."
                                    COWPER.

INTRODUCTION.

> *" Practice surpasses study ;*
> *But, if thou hast not learnt the way,*
> *Then wilt thou oft be led astray."*

HE who helps others contributes to his own happiness. Many would like to enjoy such happiness, but do not know how to secure it. To be able to help the suffering is one of the noblest gifts possessed by man, but this gift must be developed to an art, knowledge must be combined with power, if a full and beneficent result is to be obtained. However much innate talent for helping others a man may possess, success can only be secured when he knows *how* to help. In the Care of the Sick much is gained from personal experience, and by aiming at that which is practical, but at best this is a roundabout groping after knowledge, and occasionally may injure the sufferer, whilst it leads, with infinite slowness, to the object desired ; in such a case quietness and certainty in working—a powerful means for awakening the confidence of the patient—will be very late in attainment. This quietness and certainty can only be acquired by *practice* in nursing the sick; much, however, can be explained by lectures and books: the books serve for reference and for repeated reading, in order to fix in the memory what has been learnt. This is the aim of this book: it does not contain *all* that a qualified sick-nurse can and should know, but it contains *The Main Principles to be observed in the Care of the Sick.*

It teaches *what* should be done under certain circumstances; but *how* this is to be carried out in special cases, demands *personal* practical experience. Those who master these principles in their chief points *know*, not only how patients should be nursed, but how to practise sick-nursing *successfully*. It is the same with all knowledge and power. Not until practically exercising our calling are we in a position to supplement, from books, not only our knowledge, but also our power to act, and not till then are we able, correctly and clearly, to realise what we have read or heard.

Much may be said upon the qualifications necessary for sick-nursing; I will, however, lay stress upon one point: the combination and training of various qualities in the same person are more conducive to success than a highly developed single quality. The doctors and patients who observe her in her work are the best judges whether or no a woman is fitted to discharge the duties of a nurse.

In training for this difficult calling one of the first requisites is, a *very strong inclination practically to help the sick.* In herself the nurse must realise an ever-present impulse to benefit the suffering, and must feel assured that her highest happiness consists in thus actively doing good. Whether this inclination is strong enough to overcome the many difficulties and painful sensations, the dangers even, that this calling entails, will be seen in practice. If she feels happier every day in this work, and daily overcomes its unpleasantness more easily, then she should cheerfully pass through her apprenticeship. But if she has deceived herself as to the strength of her inclination— and that may happen to the best—she should choose another calling.

Women who quickly kindle to a glow in passionate excitation for the beautiful and the sublime, in whom sympathy is so strong that they experience the physical and mental pain of others in equal intensity, or imagine that they do, seldom qualify for the care of the sick, because, seeing the sufferings of others, they are crippled in their own thinking and doing. In war I have frequently experienced that persons with very tender hearts could render no assistance, from pure sympathy with, and pity for, all the misery around them ; to see others feeling deeply with him may comfort the sufferer for a few moments, but he will be more thankful to those who help him.

If women with such ardent sensitive natures can combine energy and self-control so as to compel themselves to master their feelings, still, *more quiet natures* having less to do in overcoming their feelings are better fitted for nursing the sick. The mental conditions of others not only influence us, but our dispositions likewise operate upon them: generally both are exercised unconsciously. The sick are much more susceptible and sensitive to the state of things around them than the healthy ; it is therefore evident that passionate, ardent natures, even when somewhat able to control themselves, work less pleasantly in the sick-room than those who are quiet, gentle, and patient. But there are also phlegmatic and indifferent people, slow to think and to act, who, from these very qualities, make the patient impatient, and finally become objectionable to him. Patients differ, not only in their own

characters, but different diseases produce different conditions
in them.   Personal pleasure and displeasure, sympathy and
antipathy of one towards another are called into action, and
unconsciously influence the close intercourse entered into
between patient and nurse.   The nurse must never yield
to these feelings ; in discharging the duties of her office
she must rule, however difficult it may be in particular
cases ; the doctor will guard her from exacting demands
on the part of patients, or their relatives.   Sometimes a
strict distinction is drawn between the prominent qualities
of head and heart.   True, a certain innate good-nature can
exist, even in persons of limited intellect, as a high degree
of astuteness can be combined with strong propensity to
evil.   Yet true and lasting kindness of heart always goes hand
in hand with intelligent thinking and doing; this arises
not only from occasional, transient excitation of sympathy,
but from the deep inner conviction that our own happiness is
indissolubly bound up with the happiness of our fellow-
creatures, and that, by good deeds, we not only perfect and
make ourselves happier, but we also most powerfully promote
good in others—

> " The moral system of the world is not without thee—
> It is only through thee.   Believe it, and thou helpest to make it."
>   Fr. Th. Vischer.

I must again emphasize that, more or less, a special
talent for sick-nursing must exist, if satisfactory results are to
be secured.   Natural inclination, kindness of heart, intelligence, quiet disposition must be combined with this talent: its
characteristic is, a mostly unconscious *gift of observation* of
what is occurring in, and to, man.   This is quite a special gift
—every one has eyes and ears, sense of smell, taste and feeling, and yet all are not equally conscious of things perceptible

to the senses. In children, the differences in this respect are often very distinctly marked : some notice nothing, and remember nothing, of various incidents happening around them; others observe and remember until, occasionally, one is surprised to see that they had both observed and reflected upon many things which it had never been supposed they had noticed. This power of observation may be more or less innate, but, by exercise under proper instruction, it can be more and more developed : to interpret the wishes of the patient from his eyes, and to assist the doctor in his treatment, it is indispensable. The nurse must learn to observe the patient in the medical sense, so that she may report accurately what occurs between the medical visits ; this must be learnt at the bedside of the sick.

Love of truth, sense of order, reliable fidelity in her calling, obedience to the medical instructions, *pliability* under peculiar, sometimes very uncomfortable, circumstances, are indispensable qualifications of a nurse—qualifications which must be acquired by self-training, although, depending upon the natural disposition, it may be more difficult for one than for another so to control and subordinate herself. Need I further add, that, under all circumstances, the professional nurse must maintain her respectability and morality? I scarcely think so, for whoever possesses these qualifications, and takes pains to perfect them, must be respectable and moral.

The personal qualities of heart and intellect which a good nurse should possess and cultivate, I have intentionally placed in the foreground, yet these can only be brought into full operation when existing in a *healthy body*, and that because the demands upon the physical powers of nurses are often very great : a certain wiry endurance is essential. With women, this is seldom assured before the twentieth year, and rarely lasts more than from about fifteen to twenty years; between twenty and forty years is the limit of age generally fixed for admission to Schools for Nurses. To maintain sound health is a duty of the nurse to herself, or she will soon be incapable for her calling. Nourishing food, never too much at a time, yet frequently in the day, and during night vigils, maintains the strength of the body continuously at uniform tension. A full meal, taken at once, such as the modern, feverishly excited professional life entails, in order not to interrupt the day's work, is unsuitable for women; besides, for some hours after the meal, it causes inertness and disinclination for work.

Of the utmost importance to the nurse's health, and equally to the patient's, is *the greatest cleanliness.* Frequent baths, frequent change of linen, and thorough airing of her clothes the nurse must make her duty; the exhalations and perspiration of the body settle in the clothes, and are injurious to her, and offensive to the patient. Her hands must be washed before and after doing anything to the body of the patient; especially the nails, mouth, teeth, ears, and head must be kept very clean. For a nurse, the greatest cleanliness is one of her best recommendations, as it is one of the most important means of protection against contagion, as much for the invalid, who can become infected from the nurse, as for the nurse, who may become infected from the patient. We now know that the greatest number of cases of infection have their origin, neither in the air nor in miasmata, but that the infectious matters adhere to the secretions of the patient; that these often retain power to infect, even if dried, and when, as dust and dirt in the room, they adhere to the beds, to the patient's clothes, and penetrate the skin, the mucous membranes, or the smaller or larger wounds of other persons.

On this subject I shall give special rules for guidance in particular cases.

Greater cleanliness is a principal reason why infectious diseases seldom spread so much in well-to-do families as in the dwellings of the poor; equally so, good nourishment and strengthening of the body are further reasons why infectious diseases mostly *run their course lightly, and more favourably* among those in easy circumstances. Therefore, the greatest cleanliness of the nurse must not be an object of vanity, nor an expedient to better please the patient—it is a protection for all, an *indispensable requisite* to sick-nursing in the hospital as in the palace.

To maintain the health of the nurse, besides nourishing food, fresh air and exercise are essential. After night-watching she must have some hours of rest during the day, the best times for which will depend upon the circumstances in individual cases, as will also the question whether the patient requires a second sick-nurse during these intervals, or whether some person already in the house shall take the nurse's place. If some women do wonderful things in repeated spells of night-watching, yet a woman, dedicating her whole life to the nurse's calling, would soon undermine her health and become unfitted for her vocation if she paid little or no regard to it.

On accepting the charge of a patient it is sometimes necessary to come to an understanding at once with his friends as to the hours of rest indispensable for the nurse, as well as to her food, for there are some injudicious people who will not see the need for such consideration. It is the duty of the Association sending the nurse, or of the doctor who engages her, to make the arrangements at the commencement of the nursing.

Some doctors and nurses claim to have such a *light, delicate hand* that they hurt the patients very little during operations, examinations, bandaging, &c. This would seem to be a physical quality, but when these famous hands are seen they are often neither specially fine nor soft. Then what is the cause of it? It is the skilful, sure hand, guided by experience and care, which appears light and gentle to the patient. I admit that skilfulness in giving assistance and in domestic work is more natural to one woman than to another, yet, by energetic self-training and practice under suitable instruction, much skill can be acquired. The main point always is, that, in everything she does, the nurse *thinks* of what she is doing so as not to hurt the patient—purposely of course she would not. She must also *know* how to alter the position of a patient, to put on a poultice, to give an injection, to apply a dressing without giving him pain; she may neither proceed first one way and then another in each particular case, nor ignorantly grasp him here or there. This excites and makes him discontented and cross; no patient will knowingly allow experiments to be made upon him; unavoidable pain patients will bear better if they are previously told and assured that it is necessary to their recovery but that it cannot be done without quick, transitory pain. If a patient be hurt unexpectedly he will be frightened and will scream, but if prepared for a momentary sense of keen pain, then, supposing he has been dealt with firmly but gently, and the object is gained, he will often say, "Well, it was not so severe as I had expected."

Many nurses exaggerate the care necessary in taking hold of a patient, and use only the finger-tips instead of the whole hand. For instance, if the nurse raises him in the bed by grasping both his shoulders from in front, and takes hold of him with her finger-tips only, she must dig these into his flesh in order to lift the body, and thus she must bear hard upon the patient,—whilst grasping with the whole hand causes no pain.

Something must be said of the *conduct of the nurse in her intercourse with the patient and his friends.* This intercourse is easiest with those who are seriously ill—they are mostly unconscious, or are delirious. Then she punctually and noiselessly fulfils the doctor's orders, and observes, and takes note of all changes that occur in the patient in order to report them to the doctor.

Her duty is far more difficult with those who are recovering (the convalescent), or with patients suffering from a long continuing (chronic) disease, at times feeling tolerably well, and yet requiring much assistance. In private houses such patients will often have their own nurse, specially where there is no one in the family who can or will undertake this care. However various the characters of men may be, the sick man thinks above all things, nay, almost exclusively, about himself and his illness. If he has become accustomed to his nurse and places confidence in her, then, sooner or later, he will be communicative, feeling it urgently necessary to speak of himself and his ailments. He will daily put innumerable questions, and tell her all sorts of things about his most private family affairs. If feeling weak and sleeping little, he will be excited, cross, and violent, even rude. He will cry and look despondent, will complain of the nurse, of the doctors, of the whole human race, and may fall into a rage. If feeling better, then perhaps he will praise everything just as extravagantly, will find all things beautiful, and will pour out his heart in expressions of gratitude.

The line of conduct to be pursued by the nurse in these

varying moods is extremely difficult, and experience alone will teach her what to do ; above all things she must guard against curiosity and loquacity. *Reticence, and quiet, silent, unremitting fulfilment of her calling* is her duty. If the patient wishes to speak his mind freely, she should quietly and sympathisingly listen ; she must not gossip about what she has heard, but must quickly forget all that did not relate to the illness. Whilst meeting the patient in a friendly and helpful manner, she must yet guard against becoming too intimate with him, lest she should lose his respect and he cease to obey her. For some patients it is beneficial to change nurses, as the best nurse is liable to weary from monotony in the care of a case lasting for weeks, or even months. Patients too, long confined to bed, sometimes desire a change of both nurse and doctor ; they long to break the everlasting monotony ; they hope for improvement from change, from almost anything different, even if they soon weary of it. But the nurse must discountenance in the patient any rising mistrust of the doctor ; on the contrary, she should increase his confidence to the best of her ability. No worse service can be rendered to a patient than to make him suspicious of his medical attendant. It is pardonable if, from very long-continued illness, he comes to think the doctor is not treating him properly. Medical science is not omnipotent, although most patients believe it to be so ; no patient will readily admit that his disease was born in him—each considers himself normally healthy—it was only an accident, or negligence, or wrong treatment that made him ill.

Many believe that they know the cause of their disease, even in cases where medical science for centuries has searched in vain. It is innate in man, as in the higher animals, to recognise cause and effect in all occurrences perceived by them. Now where a man observes a result the cause of which he does not know, he prefers to invent a cause rather than admit his ignorance of why this or that has happened. So with man's judgment on diseases,—each must have a cause. When healthy, it is conceivable that we cannot always recognise these causes, but if one becomes ill himself, this power of renunciation forsakes him. He broods so long over it that at last he believes he has found a cause; even a fable quiets him like a child; and if it is extraordinary in addition, he is consoled by the semblance of truth it may contain. In such circumstances little contradiction should be given. Sick people should not be contended with, unless absolutely necessary; no scientific lectures should be given about their illness; they believe only what they deem probable, and will only be charmed by instruction that confirms them in their own ideas.

Generally no invalid is of opinion that his is an incurable disease. He knows that there are such diseases, for he has seen friends and relatives succumb to them; but in his own case he seeks and finds the causes in incidental things, and not in the incurability of the disease. This is a happy thing for these patients; they hope to the last moment—this hope should not be dimmed; they deceive themselves as to their own condition, and their sanguine ideas should not be dispelled. In such cases both nurse and doctor are often placed in a difficult position. The patient presses to know the whole truth, and says, "he is prepared for everything." But this may not be quite believed; he deceives himself as to his degree of strength to hear the worst, and it becomes a duty to give comfort and ease of mind by sustaining the patient's hopes, although contrary to one's own convictions.

When we cannot cure we should alleviate suffering as much

as possible.   Hope is the best palliative, is balm to the
afflicted heart, is comfort to the despairing soul.   Yet I must
repeat, it must never be forgotten that the patient thinks
above all of himself and of his disease.   It wounds him if
his complaints, or the observations he has made, or believes
he has made, upon himself, and which he considers very
important, are disregarded : they may be quite immaterial to
the treatment of the disease, but this he must never be
allowed to notice.   Every man looks at the world from himself, and considers himself, as it were, the centre around which
all things revolve : if anything happen to him, then it
must be something peculiar.   I have always found patients
unpleasantly affected when told that their disease is quite
ordinary, one that has been seen a thousand times, its treatment proceeding according to accepted rules.   Then it seems
to the patient that he is like a cypher among a thousand
others, and he fears his case will be slightly attended to.   He
wishes to be singular, his disease, even if common, with
him should run its course exceptionally.   On one hand his
vanity is flattered, on the other, he thereby expects to receive
greater attention from the doctors and nurses.   These ideas,
which are agreeable to him, should not be dispelled.   The
nurse, therefore, must never dilate upon her experiences,
though she may say she has nursed similar cases ; but she must
never say that these cases ran their course unfavourably; and
if the patient speak of some who died from disease similar to
his, then she may tell him that most probably the cases were
different,—with him it will be otherwise.   By degrees she
becomes just as inventive in soothing replies as the patient is
in his enquiries.

These few hints point to the most difficult positions in
which nurses may be placed by patients ; much is to be

learnt from experience at the bedside. When no longer able to help herself she must ask the doctor for advice, so as not to contradict his statements. Sick people are mostly distrustful, and become all the more so if they are contradicted, or think that they are; then the idea is seized that their ailment is not correctly understood, and consequently will not be treated properly. In such difficult circumstances I advise the nurse to act as she would wish to be treated were she in the position of the patient—

> " Would'st know thyself?   Behold how others act !
> Would'st others know ?   Look into thine own heart ! "
> SCHILLER.

We will now turn from this sad picture of the care of the incurable ! Such care is a sacred duty, and not so unthankful as at the first glance it would seem to be, because the sick are grateful for every alleviation, even if it be transitory. Severe, protracted suffering compels even the strongest and most defiant gradually to contentment. It is certainly much more gratifying when, after days or weeks of great anxiety, recovery begins and advances daily towards perfect health ! The greater the anxiety that had been endured, so much the joy of the nurse is greater in having contributed her share to the happy result. The days and the nights seemed endless and full of pain : restlessly the patient threw himself about, pursued by fevered fancies. How flushed his face, how heavy his breathing, how eagerly the parched lips sipped the proffered water ! How dull were his eyes in the morning, after the excited brain had been constrained by medicaments into a short stupefied sleep ! How anxiously the mother looked upon her sick and only son, previously so blooming ! A hundred times her glance enquires, Will he live? How anxious was the wife for her husband, the support and head of the family! Silently she crept away, fatigued, and sank upon the couch, overpowered by sleep; then starting up anew, with pale features, shivering, almost numb with pain, she sees the morning dawn, and still no ray of hope, no improvement! At last, the aspect of the disease changes, the fever lessens, the nights become more quiet, a natural refreshing sleep spreads its soft wings over the patient, and he wakes as if new-born, pale and weak indeed, but with the eyes clear, the features again in their former healthy form, the voice still faint, yet already it has regained its old loved tones! And day by day the improvement continues. Like a reprieve, the tidings run through the house, Recovery! Preservation! But now *much must be done to prevent relapse*,—the food must be carefully chosen to strengthen and revive him. All rejoice as further progress is recorded. Only by degrees he realises how ill he has been; as yet he has no other wish than to eat and sleep: little by little he takes interest in what is going on around him; the past is like a lengthened dream, the details of which he only gradually recollects.

Now comes the first attempt to leave his bed, to stand, to walk again; herein he rejoices just as much as those around him. Although the attempt does not succeed so well as was hoped, for he soon feels exhausted, soon wishes to return to his bed, yet the next day he is better, and on the third is better still. At last he takes his first walk out of doors, and his strength increases rapidly from day to day. Then the nurse takes her departure, or the patient leaves the hospital. Many proofs of gratitude on the part of the patient and relatives will not be wanting, but the nurse has herself acquired the best reward: *the consciousness of duty faithfully fulfilled, the blissful feeling of having benefited a human being, of having contributed to his recovery.* The Holy Scriptures say, "It is more blessed to give than to receive" (Acts xx. 35). Treasures of gold are not necessary in order to give ; knowledge and power are often more than gold. If the nurse has properly learnt her calling she has acquired a treasure that will not be lessened, but will be increased, by experience ; to the suffering, the sick, the wounded, she can give abundantly —that is her greatest happiness : " It is more blessed to give than to receive."

The objection may possibly be raised that my demands for the qualifications of nurses are too high ; that a woman possessing so many innate and acquired good qualities will rather choose another calling than that of caring for the sick, with all its difficulties. But this objection is refuted by general experience.

I have treated sick and wounded with the aid of admirable nurses who met the highest demands made upon them. They belonged either to Catholic orders, or were Protestant deaconesses, whilst others were in no religious association but came from the Female Nurses' Schools of Germany—some of them girls and women from the highest ranks of society side by side with the daughters of citizens and artizans. No confession of faith, no class may lay sole claim to the right to learn and to help.

In some quarters it is doubted whether women in the nurses' calling can continue virtuous without the restraint of a religious association. In the name of the women themselves I must protest against such an utterly groundless idea —experience has fully proved its fallacy. To find in this protest any sentiment hostile to religion is foolish. Upon one thing, however, I must lay stress, and say it again and again in spite of the great opposition thereby evoked, that, *in a great city in which most of the hospitals are large, a female nurses' school can only be successfully developed when it is united to an infirmary specially devoted to it.*

Deaconesses' associations and religious orders have their parent-houses with infirmaries, but the directors would unanimously object to place their *probationers* for instruction in any hospital the probationers themselves might choose, though they will send *trained sisters* in groups to hospitals of the most varied character. The reasons for this will be found, not in the ecclesiastical province alone, but equally in the principles of education which have been proved a thousand times.

It is generally recognised that, in training in morals and well-doing as well as in diligence and earnestness, nothing operates so powerfully as example and habit : these cannot be replaced by any teaching, however strict. Good conduct, morality, high principles, kindness to others, cannot be learnt by rote. Train children in good company, in moral surroundings, by intercourse with high-minded benevolent people, keep all impurity far from them, and they will realize the good for themselves and see its effect in others ; they learn to appreciate their surroundings, and if the best human qualities are rooted in them, they endeavour as they grow older to create such a circle as that in which they moved in their early days. Man in his childhood is singularly susceptible to impressions from without. It is only occasionally that there are in children peculiar evil tendencies to correct by training ; upon a child growing up with good moral surroundings nothing evil from without has as yet exercised its influence ; even in later years the potency of such training still endures.

Circumstances produce much change in people in the

course of years. Untrained natures may become more noble, the defiant may learn to exercise energetic self-restraint, their self-will becomes steady perseverance in well-doing. Unfortunately, it also happens that well brought up people will change for the worse amidst evil influences; women are very inclined to conform to the tone of those around them. When a female nurses' school is combined with a hospital in which only well-trained, moral, educated women exercise their calling, the influence of the house will soon be felt by the probationers. Once a well-defined system of intercourse is established between patients and nurses, habits will be developed which, as in schools so in hospitals, will exercise a powerful, unconscious constraint. Everything seems to go of its own accord; one nurse learns from another or from the doctor, with few words, and apparently without instruction; example and habit are the principal things—they are sunshine and rain falling on fertile soil; then a rich harvest cannot fail, the weeds will soon be suppressed by the vigorous thriving of the pure seed-corn.

If the number of efficient nurses be sufficiently increased they may be sent out, not only singly to private cases, but to other hospitals, yet to these only in groups, as is done by religious associations. Such groups, small or large as may be, undertake the management and nursing of a small hospital, or the charge of a separated division of a large one. These severed branches from the vigorous parent tree can then, in good soil, raise up other female nurses' schools. Thus the institution grows, rich in blessing, branch after branch springing from the parent house on to remotest generations, until eventually every hospital will be supplied with well-trained nurses. Raise the status of the nurses, and they will gain esteem on every side. As already proved in many countries, with so noble a calling there will be no lack of girls and women for this refined position, even should the diminution by marriage be considerable. It is self-evident how important the moral element is which, in this way, will penetrate deeper and deeper among the people.

In towns and countries where Female Nurses' Schools have long been established, as in Hamburg, Baden, Prussia, Saxony, Hanover, and elsewhere, the most beneficent results have come from small beginnings. With us, too, by the aid of noble men and women, we hope to found such a school in a hospital of its own at the Rudolfiner Haus.* Everywhere these Institutions are established by Associations—

> " Combined action with the same object makes,
> Of small things great, and much of little."

Even our beginning in Vienna will be small, but the realization of its importance will spread more and more ; it should neither trouble nor discourage us if the full effect of this creation in Austria only comes to its complete development in future generations.

On entering the Vienna General Hospital and wandering through its airy, green, shady courts, we bless the provident care of the daring and progressive Joseph II., who, counselled by the eminent men surrounding his throne, raised this house " To the health and solace of sufferers" *(Saluti et solatio aegrorum.)*

The wards of even this giant hospital are now too small for Vienna ; new hospitals have sprung up around it ; a new

* This hope has been most gratifyingly fulfilled. The Rudolfiner Haus, in which about thirty female sick-nurses have been thoroughly trained, is at Döbling, Vienna. The good nursing enjoyed by patients of the Institution is one of the circumstances which spread the fame of the Rudolfiner Haus more widely, increasing the number of applicants, and thus again contributing to its success. —R. G.

science of medicine, a new system of sick-nursing exist and are ever improving. It is our duty to provide for posterity. E. Geibel's beautiful poem so exactly expresses the sentiment by which our task is animated that I cannot refrain from transferring it to these pages—

" With the old Forester at early dawn
  I wander'd through the woods, and heard the while
  The village bells ring clear, in festal peals.
  And soon the day flowed, golden, o'er the earth ;
  To their Creator's glory sang the birds ;
  'Twas just as if the very greenwood knew
  And testified the Sabbath had begun !
  Then, wand'ring still, and through plantations large,
  Where graceful saplings, by the ancient trees
  Protected, on the sunlit spaces grew,
  The old man rais'd his hand and, pointing, said—
  ' See'st thou these trees, spreading their leafy arms
  And intertwining them above our heads,
  A dense, high-arch'd, green roof, a grateful shade ?
  These are a blessing from our Forefathers !
  The things which most we need in life are all
  Provided for us by paternal love ;
  That our Posterity may have the like
  Our duty 'tis, to labour and prepare !
  When working in the forest, oft meseems
  As if with one hand I held fast the hand
  Of some progenitor, whilst my other
  Held that of some young grandchild, clasp'd close.
  So, when a sapling I would plant, my heart
  Throbs, that its very beating I can hear,

And o'er my labour must say piously
A word of quiet prayer :

      'Slender saplings,
God protect ye ! When, through the forest glades,
The Future hears the rustle of your leaves,
May Fear of God and Freedom dwell beneath
Your green encircling crowns !

       ' Posterity !
Ye then, with peaceful joy my blessing trace
As I to-day, with pious thankfulness,
The blessing of my ancestors recall !'

The hoary-headed old man ceased, and stood
As dumb in prayer, a Prophet, clear of eye,
Regarding both the Future and the Past.
Then, o'er the tender saplings round, I saw
Him, silently, his hands in blessing lift ;
But yet, among the leafy tree-tops pass'd
A whisper, like a greeting from old times ! "

CHAPTER I.

THE SICK-ROOM.

IN choosing a room intended for use by a patient as his
home for some time *the first consideration should be*, that
it can be *well aired* (ventilated); *well heated and cooled;* that
it is *accessible to the sunlight;* is *in a quiet position;* and is *not
too small.*

In middle-class dwellings in large towns these requirements
are seldom united; even among the higher classes we often
look in vain for a good sick-room. Formerly, the necessity
for making every town a fortress caused narrow and high
houses to be built; now, it is the greed for the largest possible profit. Certainly the laws on this subject have been
improved—no longer are too narrow streets nor too high
houses permitted; as much as possible, surrounding spaces
are planted with trees, yet neither Germany nor France has
attained to such developed superficial extension of towns as
England. The rapid growth of towns, and the legal limita-

tions placed upon land to be occupied by buildings rising high into the air, must compel all nations to provide quicker and cheaper means of communication so as to save time to those whose employment is in town-centres whilst their homes are at the extreme limits of the same. Then, in great towns, the middle classes will more frequently have their healthy detached houses with gardens, as in England, and in small towns in Austria.

In a detached house, with windows on all sides, a well-lighted sick-room, easy of ventilation, is more readily found than in an elegant dwelling with its windows overlooking only a narrow, dull street, and a still narrower gloomy close court.

The hospitals should be gradually removed from the centre to the suburbs of the town, and enough ground should be bought at once so that, even with obstructions caused by buildings on surrounding lands, air-movement around the hospital is not checked. Seldom has a hospital from its beginning been so well placed for its objects as the General Hospital of Vienna;—three very large courts and broad surrounding streets secure adequate light and air. Of the different styles of modern hospital buildings the pavilion (detached blocks) is the best, because it generally meets fundamental sick-room requirements, if the allotted ground is not overbuilt.

AIRING THE SICK-ROOM.

(VENTILATION.)

As man requires air to live—one of its component parts *(oxygen)* being, by inspiration, incorporated into the blood ; and another, the used portion, inimical to life *(carbonic acid)*, being exhaled,—it is clear that a man, limited to a certain quantity of air in a perfectly air-tight room, would perish, because the pure air of the room is gradually used up by him, and he poisons himself by re-inhaling air previously expelled from his lungs. This manner of death, resulting from disturbance of the respiratory action, is termed "suffocation." A man in a hermetically-closed room would suffocate slowly before he would die from starvation. Fortunately perfect air-exclusion is very difficult to secure, for the air penetrates yet more easily than water all the finest pores (to our eyes invisible) of building materials : not only through all joints and cracks, but through all building stones, through mortar, plaster, marble, &c.

But the quantity that penetrates into a room in this way is not sufficient to produce adequate *change of air* to sustain life. If there were not doors and windows which are sometimes opened, persons in a perfectly closed room built of ordinary materials and without windows and doors, in spite of this natural ventilation, would perish by degrees. When, as children, we turned our toy-boxes into homes for caterpillars, we were made to pierce many holes in the sides of the boxes, to prevent the caterpillars from dying. Man needs much more air to live than caterpillars and grubs do, that often bore holes in wood to great depths where very little air can penetrate. Airing a room, essentially signifies the constant *renewal of air.* Such renewal, where possible, should be steady and continuous. In the room itself the air ought to be, and should remain, as pure as it is outside the house, where the various atmospheric gases which surround the earth continually mix. If, then, in a closed living-room built of the usual materials the same condition of things is always maintained—the same number of persons, the same air-openings, with the same velocity of air-currents through them—the air in the room will remain the same, provided that these proportions were correct. But should a change occur, as, for instance, if a double or a tenfold number of persons be placed in the room, then, unless the air-current is increased, the air in the room will soon be vitiated. The same effect would result if, with the number of people for whom originally the ventilation was provided, the ventilation-openings were reduced. Finally, the air in the room would be affected by a change in the external temperature, or by the wind rising. Warm air is lighter than cold : the warm air rises ; below, the cold air flows after it ; by this means, without wind, a current of air is caused, often strong, and felt by us as a " draught."

From these observations it is evident, that *a good system of ventilation must admit of being regulated*—it must be, at least to a certain degree, under control as to whether more or less air shall be admitted, *i.e.*, whether a less, or a greater, motion of the air shall be produced. Further, it must be distinctly understood, effective ventilation produces what in ordinary life is termed " draught ; " the chief consideration is, so to conduct this draught that it does not strike the patients in the room, or, at least, not on any exposed parts.

"Natural ventilation," depending principally upon the difference between the temperatures indoors and out of doors, and effected by means of suitable air-openings in the room, is naturally more or less sustained as the wind is more or less ; it is therefore evident that, with perfectly equal temperature within and without, and with no wind to aid, there can be no supply of fresh air ; and that, with only a trifling difference between the inner and the outer temperatures, no rapid current can be obtained. Such conditions often exist in

the summer, but then, happily, all windows can be constantly
open. When the temperature in and out of doors is equal,
or there is only a slight difference, complete ventilation—*i.e.,*
that the air of the room is just the same as that out of doors
—can only be secured when *windows opposite to each other*
are opened; by this means air-movement can always be
produced.

Consequently I should prefer a sick-room with windows
opposite to each other to one with windows on one side only;
for, in the latter case, if even the door be opened on to the
passage (which, for a continuance, will not do in a private
house, as the passage usually serves as a thoroughfare for all in
the house) it is principally noxious air from the court, or from
the other living-rooms of the house that enters the sick-room.
In all hospitals, built partly or wholly on the corridor system,
there should be large openings above the doors of the wards,
wire-latticed only, and, opposite the doors, there should be
windows having ventilators fitted into their upper parts.

In systems of artificial ventilation ducts are provided,
having openings in the floors and walls of sick-rooms, through
which, by pumping machinery, the vitiated air is withdrawn;
or, with windmill-like apparatus, fresh air is driven in; or the
both are combined. These systems,—very costly to erect and
to manage,—have generally fulfilled their promises so imperfectly, or have borne with them so many evils, that the use of
many of them has been discontinued.

In winter, thorough ventilation cools the room so quickly
that much fuel is requisite to warm the inflowing cold air
rapidly enough to maintain equal temperature in the room.
Only cool, pure air gives real invigoration and refreshment,
yet, after experience with so-called "turret-ventilation" in
Field Hospitals, I must admit that, when the sick-room is

cooled too quickly, with low outside temperature, it is difficult
to regain the requisite warmth, even by vigorous heating.
Attempts have therefore been made to replace window-ven-
tilation in winter by conducting from without an air-channel
under the floor to the stove, surrounding the stove with a
mantle (an iron screen around the stove, and fastened to the
floor, or a screen of masonry), so that the in-flowing cold air
first passes between the stove and the mantle, then into the
sick-room, and thus is warmed before mixing with the air of
the room.   From my experience, however, this very sensible
system of ventilation cannot quite replace that by the windows,
for to do this the air-channel would have to be twice or three
times broader than usual ; but then the advantage aimed at,
the saving of fuel, would disappear.

All that I have said upon air-renewal in rooms is necessary
for healthy people in order to maintain their health, but in the
sick-room still further conditions must be considered.   Here
the ventilation must be thoroughly efficient, because pure, fresh
air is not only essential to recovery, but, in many diseases, is
the most material remedy.   The exhalations and evacuations
of the sick are often most offensive, and the question frequently arises how, rapidly, to renew the air in the sick-room
—in other words, quickly to remove the smell ; ventilation
depending upon uniform renewal of air will never accomplish
this ; it can only be done effectually by opening the windows,
and to do this in winter the heat at the same time must be
considerably increased.

Prejudices among the educated classes on the danger of
air-movement in sick-rooms have greatly lessened of late, but
as they still cling more or less firmly in some quarters I
must refer to them.

A strong prejudice exists against night air, yet, in large

towns, it is this air that is generally more pure than
the air of the day, which is polluted by dust, smoke, and
kitchen exhalations.   In towns, the windows should be open
at night to admit the fresh air.   Where this is not done, exhalation may be perceived in the bedroom in the morning,
even with the cleanest people, arising from the body itself,
and from the bedding which absorbed the perspiration of
the sleeper.   The rule, that fresh air shall be constantly
admitted into the sick-room at night, is most important
in summer; then another advantage is gained, from the
room being cooled by the night air.   In autumn and
spring when (especially at sunset) the temperature suddenly
changes, the opening for admission of air must be small.   In
winter the window should be opened occasionally, and then
very little, as the great difference between the inside and outside temperatures causes sufficient air-movement by means of
the doors and windows.   Where no special ventilation valves
exist, ample provision should be made for opening and
fixing *one of the upper parts of each window* in the position
desired.   In many good private houses, unfortunately, this
provision is lacking; for a good sick-room it is absolutely
necessary.   The cool air from the upper window-opening
falls direct, so that the stratum above the floor by the
window is the coldest in the room; here is the greatest
air-movement, and the patient, therefore, should not lie by
the window under the ventilating aperture, neither should the
nurse sit there, as her feet would soon become cold, and afterwards she would be liable to suffer from pain in them.   The
air is always hottest near the ceiling; of this any one may
convince himself by mounting a ladder in a heated room in
winter, when he will find it unbearably hot close to the ceiling, but pleasantly warm lower down.   When the ventilation

apertures are in the upper parts of the windows, or over them, the cold air enters and falls; by this means movement is produced that forces the warm air upwards and outwards. As is the strength of the air-current and the coolness of the inflowing air, so the warmth must be regulated by greater or less heating.

With the ventilation-apertures placed very high and opposite to each other, the patient lying in the lower part of the room will not be directly affected by the air-current. Yet this may happen on the door being opened. All these points the nurse must observe, and arrange the *position of the bed* accordingly; or, by large screens, she must protect it from the currents of cold air produced when the door is opened.

I cannot too often repeat, that to maintain good, pure air in the sick-room is the first necessity for the quick recovery of the patient.

It has been unanimously confirmed by all doctors and nurses serving in the late wars that, in the slightly-built, often *very* draughty, wooden huts, the wounded did not suffer from catarrhal diseases although, from lack of blankets, they felt the cold severely. The statement of Miss Nightingale that "the patient does not catch cold in bed," is, in my experience, generally quite correct; nevertheless, I must lay stress upon the fact, that much difference exists between wounded soldiers hardened against all inclemencies of the weather, and the enervated dwellers in towns. To many persons, "draught" is *disagreeable*, and, in spite of all assurance to the contrary, creates in them the fear of taking cold, and of their condition thereby becoming worse. Again, there are persons with so sensitive a skin that a draught striking on any special place will evoke pains lasting for several days; if serious illness be not thereby developed, still, to the patient, it is a discomfort from which a good nurse should save him. If patients are to be exposed for examination, for a dressing, for changing beds or clothes, the ventilators and doors must be shut, but immediately afterwards fresh air must be again admitted.

After evacuations, or after the dressing of putrid ulcers, the room must be quickly and vigorously ventilated; then quite cover up the patient, even his head, so that the draught of air does not touch him, or, when practicable, roll the bed into an adjoining room. Not to ventilate the sick-room itself but the room adjoining, is an imperfect expedient, to be used only in special cases. How ineffective such indirect ventilation is, experience quickly teaches.

If the room is very small so that the bed stands near the window (and this, if possible, should never be), then one must be contented with ventilation from the adjoining room ; but it is far better to remove the bed into the larger room, which admits of ventilation.

Here and there the opinion still prevails that *fumigations* improve the air in the sick-room, or, possibly, produce something analogous to ventilation. Fumigations with pastilles, fumigating powder, and similar articles, emit such a smoke as to render the offensive smell less observable ; yet the latter is far less injurious than the smoke, which is only imperfectly-consumed carbon and ashes floating in the room. To thoroughly fumigate a room, pour vinegar or eau-de-Cologne upon a heated fire-shovel, or let a strong spray-producer be used for a time—first with fresh water, and then with fragrant waters (those containing turpentine, such as conifer-spirit, extract of pines &c., are specially suitable).

HEATING.

Warming the sick-room (which our climate necessitates the greater part of the year), so as to maintain an uniform and comfortable temperature, is intimately connected with ventilation. I shall not here consider the various systems of Central-heating by pipes through which steam, or hot or warm water, circulates, in use in many of our large hospitals, because, over their management, the nurses have no power ; but in every case the nurse must be taught the special arrangements for regulating the temperature in the room.

As for *stoves*, one of Dutch tiles, of a size suited to the size of the room, when once heated, has the great advantage of retaining the heat for a long time, but, if quite cooled, takes a long time to re-heat. Large Dutch-tile stoves are excellent in severe winter ; but in the cool evenings, nights, and mornings of spring and autumn, when the days may be hot, it is difficult to regulate the temperature quickly by them.

In sick-rooms ordinary iron stoves are altogether bad ; they become red-hot. Even when a person is protected from the radiating heat by sheet-iron fire-screens, the air becomes dry, so that to remain in such a room is not agreeable. The "slow-combustion" stoves, of various construction, are better (Meidinger, Geburth), made internally of fire-brick, and cased in sheet-iron. These retain the heat much longer than ordinary iron stoves, and they require very little attention.*

Opinions differ as to the stoking of stoves from within or from without the sick-room. It is adduced, for the stoking from within, that the open stove draws the air from the sick-room, consuming and conducting it away with the smoke by

* H. Heim, manufacturer and patentee, Vienna ; London warehouse, 95 and 97 Oxford Street, W.—Tr.

the chimney; then, from without, the fresh air forces its way in through doors, window cracks, and pores of the walls. The open fire-place and the stove opening in the room ventilate in this way. But it appears to me that this ventilation—which can be produced by window-valve ventilation much more vigorously—does not counterbalance the disadvantages that stoking within the room carries with it.

Now whether coal, coke, wood, or charcoal be used as fuel, a quantity of dust and dirt will always be brought into the sick-room with it. After every precaution it cannot be avoided, that coal-dust and ashes eddy about the room, and that, in kindling the fire, some smoke, &c., enter the room ; and in yet greater measure is this the case with open fireplaces, which smoke when the fire is lighted, and by which, in large rooms in severe weather, the desired warmth can never be obtained unless there are stoves heated at the same time. Rooms with fireplaces cool quickly when the fire is gone out. The fireplace is a pretty ornament in a room, and one sits comfortably by a crackling fire ; but in winter, when sitting at a distant work-table, the cold is quickly felt.

How warm should the sick-room be ? No certain rule can be laid down for all cases, and it is necessary to make the following distinctions :—

1. *A patient continually lying in bed* generally requires less room-heat than a patient who is up in the day. For the first, 59° to 63° Fahrenheit ($= 12°$ to $14°$ Réaumur $= 15°$ to $17\frac{1}{2}°$ Celsius) suffice. At 59° F. ($= 12°$ R.), the temperature most agreeable to fever patients, the nurse must clothe herself more warmly, as the most comfortable temperature in our climate for the healthy is 63° to 64° F. ($= 14°$ R.); in sedentary employment the whole day long, nearly 66° F ($= 15°$ R.) are still more agreeable.

2. The temperature of *man* has defined variations, re-curring daily in the same manner. Ordinarily he is coolest between six and seven o'clock in the morning, warmest from five to six in the evening, and these differences are more observable in the sick than in the healthy,—*there is, consequently, more need of warmth early in the morning than in the evening;* in the sick-room, therefore, the fire must be maintained so that the room is not too cool in the morning.

3. *Anæmic persons need more warmth than the healthy and plethoric.* Not only moderate losses of blood, one quickly following another, or a single severe loss, cause anæmia, but in every severe, feverish, inflammatory *(acute)* disease, and in many long-continued *(chronic)* diseases, the formation of blood is checked, and is occasionally suspended for a time. For such persons the temperature must be higher than previously stated. *As the doctor directs, it must be raised* from 66° to 70° F. (= 15° to 17° R. = 18·75° to 21° C.); during lengthy operations, the operation-room must be heated even to 77° F. (= 20° R. = 25° C.). After severe, sudden losses of blood, the temperature of the patient sinks so rapidly that it is often necessary to completely wrap the whole body in warmed blankets; by this means the fleeting life is sometimes retained, especially when persons are lying at the same time in a deep faint and cannot swallow and nothing can be administered internally.

4. Although the temperature in the room is cool, very sensitive, excitable people, and patients suffering from heart disease, occasionally have a transitory sensation of heat, and they fling open all the windows—the next moment they are again cold; for them it is always either too hot or too cold; if their temperature be taken by the thermometer, no foundation is found to exist for these sensations. The nurse must

learn to distinguish whether she has to do with such easily excitable people, or with feverish people, with whom sensations of heat and cold often alternate. To the first, she must not yield by causing continual alternation of more and less ventilation, of more and less heating, but by other clothing, lighter or heavier blankets, and hot bottles, she must meet their wishes ; but some are never to be satisfied, and, in such cases, she must have patience, must persuade, and pacify.

5. Although daily regular heating of the sick-room does not form part of the nurse's ordinary work, *yet*, paying strict regard to the thermometer, *it is her duty to regulate the temperature* by adding fuel to the fire, or by opening the ventilators.

6. Every sick-room should have an out-door thermometer ; in the morning the amount of fuel requisite for the fire must be determined by the state of the outer temperature.

COOLING.

No less important than heating in winter is the cooling of the sick-room in summer,—then it is necessary, above everything, to keep the sick-room windows open all night long, and partially to close them towards eight o'clock in the morning. Where rooms are exposed to the sun, Venetian blinds should be fixed : those are the best, like the Marquise, which can be pushed outwards, and kept in this position by rods. The room may be cooled by placing therein large blocks of ice in shallow tubs, so that the ice-surface is as large as possible. Spraying water and eau de Cologne does much to cool the air, and will be refreshing. As the effect of "spraying" is not lasting, it must be often repeated, yet not too frequently with eau de Cologne, because, if used in large quantities, it acts as a narcotic. Lastly, large branches of trees in a vessel of water in the room contribute to the coolness, as will also large wet sheets hung before the open windows.

SUNLIGHT.

A sick-room ought to be *well lighted*, not only that the changes that occur in the patient may be observed, but because light is essential to man. We see that persons who have little light whilst at their work, especially miners, always look pale and sickly,—this etiolation from want of light is particularly observable in poor, wan children, who grow up in dark underground dwellings,—but we may see the influence of light still more forcibly exhibited in its effects upon plants. The shoots that frequently grow from the potatoes in the cellar; the oats, sown indoors in winter on wadding saturated with water, all spring up pale and almost colourless. All creatures that live in subterranean waters, as the proteus in the Grotto at Adelsberg, are colourless; if brought to the light, gradually they become brown or black. But even with the sunlight there must be special conditions: all sorts of trees, shrubs, and flowers may be planted in a garden surrounded by houses; if properly seen to, many will thrive well for a time, but few will flower, and that only in situations that admit of the sun shining daily for some hours upon them. Grass, which at first grew well under large spreading trees, gradually withers from want of sunlight; with much moisture, it will be supplanted by moss, which needs no sunlight. Flowers turn their faces to the sun —generally a shady garden will not yield fruit. The little trees, growing untended in the woods in the sunlight, will resist the severest cold of winter much better than similar trees planted in the confined town garden, with little sunlight. Thus the sun operates, not only upon the development, blooming, and fruit-forming powers of plants, but it makes them stronger and more capable of resistance to external influences. And although we cannot measure, accurately in every case, the influence of sunlight upon man, yet of its existence there is no doubt; it is not only the cheering effect of the brilliant blue heavens, and of the sunlight, which the patient requires—it is of the greatest importance for his recovery and invigoration that the sun shall shine into his sick-room.

Only in few diseases must the light, specially the sunlight, be excluded; for instance, in some eye-affections, this is necessary, as also with some nervous patients who, like those suffering from hydrophobia, fall into painful convulsions at the sight of shining surfaces.

SIZE OF THE SICK-ROOM.

For hospitals, certain principles have been established as to the amount of air-space every patient should have, in order, with moderate ventilation, not to suffer injury from vitiated air, nor to cause it to others in the same room. Accordingly, from 46 to 52 cubic yards of air should be allowed for each patient.*

If we have, for instance, a room 10 yards long by 6 broad (the floor surface will be $6 \times 10 = 60$ square yards) and 4 yards high, then the room contains (4×60) 240 cubic yards of air, and, therefore, from five to six beds can be placed in

---

* In the wards of my clinic 47 cubic yards are provided for each patient, including the two attendant nurses. The beds are not always all occupied.

it. This may be taken as the general rule, and is a guide for choosing a sick-room in private houses.

. The shape of the room must also be considered, still more the provision for adequate ventilation. Formerly, churches were preferred for military hospitals ; from the internal height of many of them there were often more than 130 cubic yards for each of the wounded, even when the beds were close to each other, yet these were churches always offensive. There were no possible means of ventilation, for air-motion in the highest part there was none, or it was so insignificant that the exhalations from the wounded and from their uniforms, knapsacks, &c., could not be overcome.

Too large a room for a single patient is dismal, and expensive to warm ; too small, is difficult to ventilate so that the patient is not exposed to strong currents of air, and further, it acts upon him oppressively. One of medium size is the best.

QUIET POSITION OF THE SICK-ROOM.

For feverish and nervous patients nothing is more distressing and exciting than being obliged to hear the din of the street all day long, and partly through the night, or many persons frequently passing in the vicinity of their rooms, or in an adjoining corridor, or overhead. They restlessly start up, and this often recurring, they tremble and begin to moan. Music, hammering, rattling of carriages, or other noises in their neighbourhood are highly exciting to them, shorten their sleep, or prevent them from sleeping at all. The effort should therefore be made to find a room into which such noises do not penetrate.

CLEANSING, AND FURNITURE OF THE SICK-ROOM.

Keeping the sick-room clean is of the greatest importance. Air, light, and warmth are necessary for the welfare of man, yet we now know that infectious matters are not atmospheric gases, that they are not soluble in liquids, but that they are most minute bodies (seeds or so-called spores of the minutest fungi) which float in the air or in liquids, and settle on, or obtain entrance into, the body with dust or with liquids. Dust, therefore, must be banished from the sick-room as far as possible. This is not an easy matter in comfortable private houses, because the rooms are generally so arranged that many things are in them (as, for instance, curtains, carpets, furniture), from which the dust can be removed with difficulty.

Hospital wards should be so arranged that dust not only rests with difficulty, but is easily removable ; and this is attained as follows : all surfaces of ceilings, walls, and floors must be as smooth as possible. The choicest material would be marble in slabs, or artificial marble, or, next to these, glazed earthenware tiles ; for the floors, mosaic or asphalt. With us these materials are not taken into account, because (in Austria and Germany) they are too costly, and the stone and smooth floors are rather cold if hot pipes do not run beneath them. Certainly in winter, carpets may be laid down (as in France and Italy), and nurses and patients can wear felt shoes. But these articles are dust-catchers. Experience has not yet determined whether washable, smooth india rubber or linoleum floor-coverings suffice to keep off the cold of a stone-floor, or whether these materials are so durable as to compensate for the expense of providing them.

Stone-work as facing for ceilings, walls, and floors is, in its first cost, very expensive, but is very durable, and requires no renewal. The best substitute is, painting with oil colours and then varnishing. For the first few years this painting must be renewed, specially on the floor, but by degrees the wood will be so saturated with colour and varnish that the surface becomes as hard as stone, and small repairs will be needed only at those places where the traffic has been great.

The objection has been made that, by painting with oil colours, ventilation by the pores of the walls will be quite suspended. I have already referred to this; as a means for changing the air in the sick-room in spring, summer, and autumn, I do not estimate this source highly, even if a storm blows against the otherwise perfectly closed walls.

If the walls are whitewashed, the washing must be renewed at least twice yearly. To decorate them every time would be waste of money. In a room permanently devoted to the reception of patients, wall-papers must be rejected; the same with parquetted floors, which require much polishing with wax, and may not be washed.

For walls, ceiling, floor, and *furniture in a sick-room* it is most important that everything can be *washed and wiped with damp cloths* without being injured; under ordinary circumstances, dusting would have to be done several times a day, with the window open, and the carpets and curtains would have to be removed and beaten every morning to keep them as free from dust as possible. Consequently, all sick-room furniture—night-table, chairs, wardrobes—should be painted in oil colours, inside and out, so that they may be easily washed. We shall return to this subject when dealing with the Disinfection of a Sick-room, and shall show how easily and quickly such a room can be cleansed and disinfected, and how difficult it is to do so when the room is hung with paper, the ceiling beautifully decorated, the furniture upholstered, the floor carpeted, and the beds, windows, and doors curtained—indeed, it is impossible to get the finest dust out of these articles without handling them roughly, or destroying them.

How many of these requirements of a sanitary sick-room are to be obtained in a private house must be left to the doctor and nurse; the effort should be made to secure as many as possible. It is one of the chief duties of a nurse to keep the sick-room clean; but it is not judicious to take up her time unnecessarily with beating carpets, as this is prejudicial to the care she should devote to the patient.

It is not wise to clean the sick-room very early in the morning—many patients only then fall asleep, and they should not be disturbed. On the patient waking, the nurse should give him his breakfast, make his toilet, and, where possible, remove him to another bed (if in an adjoining room, all the better); then the necessary cleaning of the room can be accomplished without annoyance to him.

THE SICK-BED.

Who, indeed, has not occupied a miserable bed? One tosses about and cannot fall asleep, and, with more or less reason, the whole blame is laid on the bed. "His bed is well-made" is aptly said of anyone who is in very comfortable circumstances.

In hospitals, everything ought to be arranged in the best manner for the sick—particular attention must be given to bedsteads and bedding. The chief thing is, that the bed shall be comfortable and suited to the patient, yet, at the same time, that it shall be made easy for the doctor and the nurse to support, to bandage, and to wait upon him in the bed, as this will ensure the least discomfort. Modern hospitals have only iron bedsteads fitted with independent woven-wire spring mattresses.

The iron bedstead has this great advantage—it is most easy to keep clean, and, suitably constructed, is the most easy to move; it should have no joints in which vermin can hide. It took a long time to get accustomed to them—they were too airy, too cool, and ugly, because one could see through them everywhere. That a *bedstead* serves at all for the warming of a patient is an error. An hospital iron bedstead is not beautiful; if ornamented on the front and sides with sheet-iron, painted and japanned, the disadvantages again arise as with the wooden bedstead, viz. :— a number of uncontrollable corners and crevices are formed. For whom should the bed look beautiful? It can only be for those around; the invalid in the bed sees nothing of it, and yet *it is he* for whom care must be taken. Shall he be bitten by bugs, which creep into the joints of the bedstead and cannot be removed unless it is taken to pieces and brushed over with offensive tinctures, because those around him do not find his bedstead beautiful?

The length, breadth, and height of the bedstead are of importance. For an adult it should be about 6 feet 6 in. long, and about 40 inches broad; greater length is seldom necessary —greater breadth never. For the patient's getting in and out, sofa-height is most comfortable (about 20 inches), but it is very inconvenient to wait upon him in so low a bed, to examine or to bandage him; for a permanency, it is not to be endured by doctor or nurse. To lift a person from this height, for instance, to change beds, or gently to lay him down again so low, requires considerable strength; the level at which the patient should lie is from 30 to 33 inches.

The legs of the bedstead must have *castors* (piano-castors) sufficiently large to admit of the position of the bed being easily changed. Castors at the head-end will suffice, as the foot-end can be easily lifted even with the patient in it, and, with skill, can be easily guided. Bedsteads with four castors are objectionable, as they move too easily and yield to every slight pressure.

Box-framed steel-spring mattresses, with the springs as in upholstered furniture, are generally used. These are covered with close-textured linen or cotton material, under which covering a layer of horse-hair or wool is sometimes placed. All this should be avoided, as it may absorb infectious matter, is liable to be soiled, and must be frequently washed. This is very inconvenient, as the material must be removed for washing and must be replaced, entailing unnecessary trouble and cost : labour is money and time. The now much-used "wood lath spring-mattresses" are more practical and are inexpensive—in these, long, thin laths rest on spiral springs and form the underlay to the mattress. The "woven wire spring-mattress," consisting of a strong frame over which a kind of webbing of spiral springs is stretched, has surpassed all other articles for this purpose. In an emergency, in war for instance, a good box-mattress can be quickly made by tightly stretching a close-textured material (ticking, &c.) over a firm wooden frame fitting into the bedstead.

As it is unavoidable that every bed yields to continuous use, a wire spring mattress must be tightened anew from time to time, and be fresh painted, or be replaced. It is also right to say that its tension, calculated for a medium weight of body, is not equally comfortable for every one—a heavy man sinks into it, and it is hard for a thin, light woman. It is the same with every description of bedding. The filling of palliasses, and the stuffing of mattresses are generally calculated for medium weights.

Palliasses as under-mattresses, or even as the only couch, cannot be avoided in war and in practice among the poor, but they make bad, hard beds; if the straw become wet it rots, and is offensive—vermin will swarm in it, and it is never without dust.

For *top mattresses* (the mattress upon which the body lies) all kinds of materials are used—hay, seaweed, cotton, jute, "ligneous fibre" (fine wood shavings), &c., &c.

No material as yet has taken the place of prepared *horse-hair.* This is the most expensive part of the bed, not only at first, but in its subsequent treatment; the mattress will unavoidably be soiled occasionally, and then it must be taken to pieces, washed, aired, and again stuffed, at much labour and cost. The mattress must be stuffed fuller in the centre than at the sides, or a depression is quickly formed. To stuff a mattress so that it shall not be too hard to lie upon, nor be easily compressed, requires much practice. Tying down or tufting the mattress is an additional, but very necessary labour, to hinder the hair from being too rapidly compressed and forced to the sides. It should be from 4 to 5 inches thick at the edges, and 5 to 6 inches in the middle. In many cases it is better to make the mattress in three parts, crosswise,—for surgical patients this is sometimes necessary, when certain apparatus have to be applied. Thus divided they are more durable, because their position can be changed; if in one piece, the same part is always subject to the heaviest burden; the horse-hair mattress should not be laid directly upon the wire mattress, lest its cover should soon be chafed through; ticking laid between gives adequate protection.

The *sheet* is spread over the mattress; it must be white, and of sufficient size to tuck under the edges. To keep the mattress from being soiled, insert a waterproof sheet between mattress and sheet.

Pillows should be of horse-hair only, not too firmly filled, and encased in white linen. Many beds have a wedge-shaped bolster at their head. However practical this may be, it has certain disadvantages. It must have a specially-shaped case —it is too high for some patients, too low for others.

I believe the arrangement of beds in my clinic to be very practical. Each has two pillows as wide as the bed, about 15 inches deep, and nearly 8 inches thick in the middle. They may be placed together by the patient or the nurse, as is most comfortable to him. Each ward has a reserve of pillows, to be used as required.

I consider *feather beds* and *pillows* very bad for patients, as the body sinks in them and becomes unnecessarily heated; for the head, feather pillows are particularly injurious, as with them the head is nearly always perspiring, and the scalp becomes so sensitive that, when the patient begins to get up, he must wear thick caps for a time to avoid pain. Such pillows, saturated with perspiration, are difficult to cleanse, to dry and air thoroughly—if this be not frequently done a disagreeable musty odour ensues.

For *covering*, one, two, even three blankets in winter—a light cotton coverlet in summer—are generally recommended. These must lie upon a white linen sheet, which is turned over to the outside on to the blankets, and there fastened. In hospitals, top sheets are the same size as the under, and the pillows and cases are all of the same size, which much simplifies the distribution of the linen. Folded sheets are used as *underlays* in special cases. As a matter of course, all bedding must be *perfectly aired*, and must not be too cold when used.

In private houses the difficulties will not be trifling which the nurse will encounter in introducing these principles for arranging a sick-bed suited to its purpose, and in simplifying the furnishing of the sick-room; she must be careful not to put the patient and his surroundings in a bad humour by too many innovations at once. In infectious diseases she will best succeed by telling his friends that the infectious matter can settle in the feather-beds, pillows, carpets, curtains, &c. in the room, that these articles will materially suffer subsequently from disinfection, and possibly will have to be burnt. It will be difficult to persuade enervated patients to dispense with plumeaux or eiderdown coverlets, although these are less injurious; they can be replaced by blankets, but these weigh more heavily—in such cases, one must sometimes yield to habit. Habit! How we all cling to it!—it is difficult to deprive one's self of anything, even when perfectly conscious that its retention is injurious.

Some things that will improve the position of the patient when lying down must still be mentioned.

When the upper part of the body is in a moderately high position many prefer to have the head high. Then the small bolster head-rest, not too thick, stuffed moderately firm with horse-hair, or *hard* with feathers, is the best for the purpose (when *hard*-stuffed with feathers, the beforenamed disadvantages are lessened, and the great elasticity is agreeable); this head-rest should be provided with a cord, fastened to both ends of the bolster, so as to admit of its being slung over the head of the bed, or over the shoulders of the patient. It gives relief and change to the position of the head, and is cooling to the head and nape of the neck when the situation of such a head-rest can be frequently altered. By using this sling-bolster head-rest the change of position is more simply and more easily effected by the patient himself than by changing the pillows, which frequently necessitates his sitting up in the bed.

Patients who prefer to lie with the whole upper part of the body raised, or who, from difficulty in breathing, must lie high, and corpulent heavy persons lying in a moderately high position, should always have a pillow under the loins—without this they lie hollow, and suffer from lumbago. Those who lie high, and heavy, or feeble, moderately high-lying patients, *slip* easily, and always more and more *downwards to the foot of the bed*, often getting into a miserable condition. Consequently, to help feeble patients, the upper part of the body must be placed in a somewhat lower position in the bed, and the whole body frequently lifted upwards: in so lifting, the body must be laid hold of at the hips, *not at the shoulders*. More will be said on this point in the chapter on Sitting Up to Eat, Read, &c. Should the patient be tolerably strong, then a block of wood, nearly the width of the bed, and bevelled on the side towards the feet, may be laid at the foot of the bed so that he may push against it. Failing such contrivance, a footstool with the upper surface towards the feet, or a small drawer, say from a night-table, may be substituted. To patients not strong enough such arrangements are useless,—they must be, again and again, drawn upwards in the bed.

Persons sitting up in bed for some time are better supported at the back by a bed-rest, (a specially-made adjustable wooden frame with webbing stretched over it), than by many pillows, which constantly shift. With *very restless patients* one must take care lest they fall out of bed in an unguarded moment. To prevent this, planks, sliding panels, or bed-staves can be inserted at the sides of the bed, specially at the middle part, such as are often used when little children sleep in large beds.

It is not possible to describe without illustrations every little help that can be given to maintain the position of the patient in bed,—that must be learnt practically.

It is of great importance to select the most suitable *position in the room for the bed*—to gain space, beds are usually placed sideways against the wall. When nursing a patient continuously confined to his bed, this is extremely uncomfortable for the nurse, and consequently for the patient. The bed must stand free, so that he may be examined and attended to from both sides. If the room be too small for the bed to stand quite free, then place the bed-head against the wall. The fire (or stove) must not be too near the bed, lest the radiating heat should strike the patient on one side ; neither should the bed stand just in the direct line of the ventilating current, unless such current is high up in the room.

It is very pleasant, especially in convalescence, when, from his bed, the patient can see the green trees through the windows, or at least something of the sky. Temporary dazzling by a too brilliant blue sky may be lessened by Venetian blinds.

CHAPTER II.

GENERAL RULES FOR THE CARE OF PATIENTS CONFINED TO THEIR BEDS.

CARE FOR GOOD POSITION AND COMFORT OF PATIENT IN BED. SMOOTHING SHEETS, CHANGING AND SHAKING-UP PILLOWS. BED CRANE; BED TRACES. CHANGING BODY-LINEN. CLOTHING IN BED. EVACUATIONS IN BED. CHANGE OF BED—CHANGING FROM ONE BED TO ANOTHER. CHANGING BED-LINEN. UNDERLAYS: WATER-PROOF MATERIALS. WARMING THE BED.

BED-SORES AND GANGRENOUS DECUBITUS. BED-SORES: PREVENTIVE MEASURES; CIRCULAR-CENTRE CUSHIONS; LOTIONS; PLASTERS. GANGRENOUS DECUBITUS: CAUSES. WATER-CUSHION: ITS USE. CLEANSING THE DECUBITUS. TREATMENT OF THE WOUND.

LIGHT AT NIGHT IN THE SICK-ROOM. LARGE CLOCK WITHOUT STRIKING ACTION. DRINKING, EATING, READING IN BED. SITTING UP AND DRAWING-UP OF PATIENTS IN BED.

IMPOSSIBILITY OF EFFICIENTLY TREATING AND NURSING THE POOR IN THEIR DWELLINGS.

EXCELLENT OBSERVATIONS AND REMARKS OF MISS FLORENCE NIGHTINGALE UPON THE PECULIARITIES OF MANY PATIENTS, AND UPON THE CONDUCT OF THE NURSES.

M ANIFOLD as are the diseases in which nurses have to assist, there are yet certain rules which serve for all patients obliged to lie for a long time in bed, and these rules we will now consider.

For giving relief in continuous lying in bed, and for the prevention of bed-sores, many practical measures have been approved.

Above all, *the sheet must be carefully put on and be firmly fixed, so that no creases are made.* Several times daily it must

E

be drawn smooth, and the creases removed by the hand. If
with little effort the patient can raise himself somewhat (by
the bed-crane, see page 67) then *one* person can easily draw
the sheet smooth, first on one, then on the other, side of the
bed. If too weak to lift himself, *two* persons must draw
the sheet tight from both sides at the same time, and then
from the head and from the foot of the bed. This must be
done slowly and equally several times in succession. The
sheets must be firm and strong—very fine or old sheets would
be torn.

The *shirt* also must be often drawn smooth on the back ;
creases in the shirt as well as in the sheet favour the formation
of bed-sores.

Very refreshing to the sick is *the changing and shaking-up
of the pillows*; they not only get pressed together, but they
become warm and damp from perspiration — this specially
applies to pillows placed under the loins.

When a patient lies for a long time on his back because he
is too weak to turn himself round, he often suffers from heat
in the back. *To cool the back*, lay him carefully upon his side,
keep him so for a time, and draw his shirt smooth ; or, if the
doctor permit, he may sit awhile, during which time the back
and head-pillows may be re-arranged.

These changes can be easily made if the patient himself
can help a little, and, for this purpose, *bed-cranes* and *bed-
traces* render excellent service—in many cases they render
the assistance of a second person unnecessary.

Cords with handles hanging from the ceiling are not
practical, because they get beyond the reach of the patient
when the position of the bed is changed—which change, in
some circumstances, is very desirable. I find very useful a
leather strap suspended at the end of an arched iron curve,

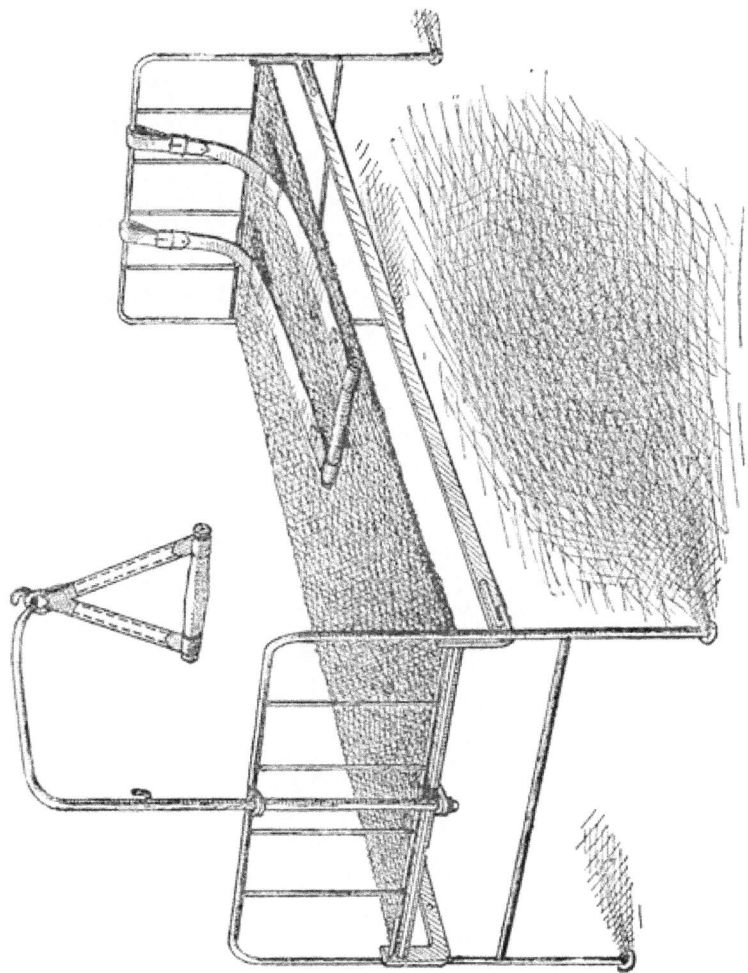

THE BILLROTH BED-CRANE, BED-TRACES, AND BEDSTEAD.

and hanging down towards the shoulders of the patient. This crane must be fixed to the head of the bed by its long upright part, and must be strong enough for the patient to pull upon it with the full weight of his body. Grasping the handle (the strap of which can be shortened or lengthened at will) with one or both hands, he can raise the whole upper part of his body with little effort, and, by pressure of the heels, the whole pelvis too as much as is necessary. This contrivance I had made for my patients in Zürich, and brought with me to Vienna. In the field hospital, as a substitute, I had a divisible wooden crossbeam made and set up transversely over the upper part of the body of the patient ; to it, the leather-strap with handle was fixed. Any carpenter and saddler can make such a contrivance in a few hours. But I prefer the crane at the head of the bed, because it is less in the way when assisting or bandaging the patient, and the cross-beam must generally be pushed aside when the upper part of the patient is being attended to. These cranes do not render bed-traces unnecessary in every case. In bandaging the head or chest the doctor can be much hindered if the patient support himself by holding his arms high up to the crane. In that case, it is better to fasten round the foot of the bed a girth long enough to reach to the middle of the body ; this can be shortened at will, and at its upper end it must have a loop or handle so that he may maintain himself with less exertion than when holding the girth tightly with both hands. In emergency any one can arrange such bed-traces by tying two towels in a large loop around the foot of the bed.

Change of body-linen and *washing of the body*, the *back in particular*, are refreshing to, and very important for, patients long confined to bed, especially for those who perspire freely. In washing, the bed must not be wetted, and the ventilators must be closed—temperature ranging from about 66° to 70° F. (= 15° to 17° R. = 18° to 21° C.). The water must be warm (about 95° F. = 28° R. = 35° C.) when cooling ablutions are not ordered by the doctor; the patient should be dried with a coarse soft Turkish towel. If the whole body is to be washed, it must be done gradually: first, the head, face, and neck, then chest, and abdomen, then arms, and one leg after the other. *To take off the shirt in bed*, a certain measure of dexterity and practice is necessary: the patient is first raised a little, or he raises himself, so that the shirt can be taken from under him and be drawn upwards. If he has pain in one side of the head, in the throat, the chest, or in one arm, then the arm of the healthy side must first be lifted, and the sleeve drawn off whilst the arm is drawn back: strip the shirt over the head to the sick side, and slowly and carefully draw it off the arm. If the arm be bandaged so that the shirt sleeve can only be removed with difficulty, the patient should not be daily troubled by such removal, but the sleeve should be unstitched. Putting on the clean shirt necessitates the same movements but in reverse order: first drawing the sleeve over the affected arm, then passing the shirt over the head, then introducing the healthy arm into its sleeve, and then drawing the shirt down under the back so that it lies without creases. If raising the patient to change the shirt must be avoided, cut it open its whole length at the back, so that it can be put on like a pinafore with sleeves.

If the arm of the sick or wounded person, or of the person who has been operated upon, is fastened by the bandage to the chest, then, in drawing on the shirt, first the free arm and then the head are introduced through the shirt holes, and the shirt is then drawn down and fastened at the neck. The empty sleeve must be fixed in front, otherwise it is liable to be caught with the sheet under the pillows, or under the back, and so the bandage would be drawn upon with every movement of the patient.

As *clothing in bed*, men need nothing but a shirt; only those to whom a low room-temperature is disagreeable, or who prefer to lie with the upper part of the body exposed, should wear a woollen night-jacket. A bald-headed person, especially if accustomed to a wig (most uncomfortable to wear in a sick-bed) should put on a nightcap. When chemises are low-necked, women wear night-jackets in bed — many use nightcaps.

Nurses must give special attention to women's hair. Here I am not taking into account the urgent necessity carefully to cleanse and free all patients from vermin before they are laid in the hospital. The long hair of women must be braided in two plaits up to the head, and these laid in front upon the chest, or fastened on the top of the head; at all events the hair must be dressed before changing the body-linen, and prior to bandaging or changing beds. If this be not early attended to, the hair, unplaited, or not plaited up to the head, gets so entangled that, subsequently, without giving much pain it can scarcely be disentangled. With patience, disentanglement is more easily accomplished by using a hair-brush than by a comb, particularly after oil has been rubbed into the hair.

As we are dealing with the toilet it may be at once stated, of the sick as of the healthy, that the *face* and *hands* must be *washed*, and the *mouth*, *ears*, and *nails cleaned daily;* if not, the patient becomes dirty, and it will be more difficult afterwards to cleanse him without taxing his strength.

For evacuations in bed, as well as for the reservation of expectoration, special vessels are made for men and for women, the practical use of which can only be demonstrated at the bedside. Bed-cranes are here of great value. In providing the vessels referred to, only those of china or of glass are fit for use ; care must be taken that the covers fit closely, and are without grooves or unnecessary cavities, so that every time used, they can be quickly and completely cleansed without difficulty. In no case must the evacuations remain in the sick-room, but they must be taken at once to the water-closet (except the expectoration vessels). Pewter or tin bed-pans, upon which a padded ring is laid, must never be used. In winter, before use, the clean empty bed-pan must be rinsed with warm water, or its coldness will affect the patient unpleasantly.

To many patients, frequent *change of bed* is of the greatest consequence. For those seriously ill, two beds should always be prepared ; the change not only refreshes, but it most certainly protects against bed-sores.

When these instructions have been very carefully and rigidly attended to, the sufferer will be greatly benefited. If the same things be unskilfully done they cause agony and danger to him. It is incredible how much human awkwardness is made manifest when a patient's bed is changed—as much ineptitude of mind as of body. Most persons think the removal of a patient from one bed to another is only a question of personal strength, but, after some instruction in deftly handling him, they will acknowledge that strength is not the chief requisite.

A patient *is removed from one bed to another best and most carefully by one person.* To make the distance as short as possible, place the head of the empty bed at the foot of the one occupied ; then stand on the right side of the recumbent patient (on the left side when the painful parts are on the right) ; bend the knees, bring the right arm (after the shirt has been drawn well down and the coverlet removed) as far as possible under him, so that the upper part of his thighs rests upon it. Then pass the left arm quite under the middle part of the back of the patient to the other side ; tell him to embrace your neck with both arms, not to stiffen his legs, but to leave them quietly hanging. Now lift him whilst you straighten your own bent knees, and rise, bending your own spine backwards until the body of the patient lies upon your chest. Then carry him to the new bed and gently lay him in the middle of it, whilst you carry out all previous movements, but in their reverse order.

It is more difficult to lay the body gently down than to lift it ; therefore it is advisable for a second person to stand at the side of the bed opposite to you, from thence to lay his arms also under the patient and, to a certain extent, to receive and assist in laying him down.

After some trials, begun with light, healthy persons, it is surprising what heavy people one is thus gradually able to carry. It is *upon the chest* that the weight must be carried by means of the strong muscles of the back and neck. To lift, carry, and gently lay only a moderately heavy person in the bed, *with the arms free*, needs considerable strength, not often found in doctors and nurses. Should the patient be unconscious, and be unable to hold up his head, or to clasp with his arms, then his head must be held by a second person. When the foot is injured it must be specially supported by a nurse, and be kept from hanging down and from knocking against anything. On the whole, I always prefer to take hold of a patient from his right-hand side, because the stronger right arm more securely holds the heavier lower part of the body to be carried.

On the bed being changed, all bedding from the former

bed must be immediately removed, so as to clean and air it. Should the room be too small to allow both beds to stand with their narrow ends together, they should be so placed that the bearer has not far to go to the new bed: above all he must remember, that he must come to the same side of the new bed as that from whence he bore the patient— in changing beds, want of attention to this point leads to perplexity and confusion.

Although I am satisfied that many nurses, by practice, could carry a moderately heavy patient, yet, for very heavy bodies, it is necessary for two nurses to practise together *the carrying by two*, and for this the following rules should be observed: the two bearers must take hold of the patient on the same side, one placing his arms under the back, whilst the patient clasps him around his neck; the other and stronger bearer placing his arms under the pelvis and thighs. The lifting, walking, and laying down must be done simultaneously by both—is best done by word of command; the carrying upon the chest must be as previously described. If the two bearers feel that they can scarcely manage the weight, then a third must carry the head and a fourth the legs, but all must take hold of the body on the same side, or the most painful dilemmas will arise on laying him in the new bed; these would be almost ludicrous were the matter not so serious. Removal by several persons acting together must be specially and accurately practised, as such removal is far more difficult than when done by one person alone. To embrace the patient from in front in order to join hands at his back is wrong; it will soon be seen that considerable strength is needed, even with two, thus to lift, to carry, and to lay a person down; in this way the poor patient, awkwardly grasped, can only be thrown with a jerk into the new bed.

In the practical drilling exercises necessary for this purpose the importance of a proper height for the bed, as previously stated, will be immediately recognised. If the bed be too low, very strong muscular powers will be needed to rise quietly from the deep bent-knee position, with the body in the arms, because, in a low, kneeling position, it is difficult for the bearer to bend his body far enough back so as to lift the patient on to his chest. If a person lying upon the floor has to be lifted, it usually takes place as follows :—one lays hold of him by the arms, in favourable cases at the shoulders, whilst another takes him by the legs, and then, with a swing, the man is thrown upon a sofa or into the bed. Apart from the fact that it is frightful to see, it is most injudicious. Kneel right down beside the recumbent person, slide the arms under him as beforenamed, and then rise slowly, whilst, standing opposite, a second person grasps your wrists and renders every support in his power,—then carry as described. With very heavy persons, two bearers on each side are necessary.

It is an error to suppose that the body of the woman is less qualified for carrying than that of the man. In Austria and Hungary, as in Italy, it is principally women who carry heavy loads up the scaffoldings to the masons. In Italy I have seen women carry heavy loads up and down hill; they bore themselves much more skilfully than the men, who evidently considered carrying no portion of their work. One wonders at the unusual development of strength some women have attained. It is wrong to say that carrying injures healthy women; there is no reason why a healthy female nurse should not learn to carry a patient as easily as a male nurse does. Till recently, I not only personally lifted from the operation table into the bed a great number of those upon whom I had operated, but daily I myself changed them from one bed to another, and I insist upon my assistants learning to do the same ; they are soon convinced it is more a matter of practice than of special bodily strength. In hospitals, it is necessary that, from time to time, the doctor should perform such services at the sick-bed, so that the nurses shall see first, that no service rendered to the sick is to be deemed insignificant or beneath their dignity ; and secondly, that even in all seemingly most insignificant trifles the doctor always knows and can do more than they, and therefore has the right to instruct them.

If the patients are so weak that it is feared changing to another bed would injure them (a very rare case), or if, with wounded patients, or with those who have been operated on, the removal is difficult, then, *how to change the under-sheet without requiring much movement of the patient must be fully understood.*

For this purpose, from one side of the bed the sheet is rolled up lengthwise, close to the body of the patient ; then the clean sheet is also rolled up lengthwise almost to the middle of its breadth, and this roll is laid by the side of the roll of the used sheet ; whilst the patient is lifted, both rolls are quickly pushed under his body. The used sheet is removed, and the clean comes into its proper position and is unrolled and fixed. If the patient can support himself for a moment by a bed-crane, one nurse is enough to change the sheet.

If strong enough to lean upon his arms and to lift his body a little, so as to shift his position by pushing with his feet, the change of bed can be made as follows : the two beds are placed side by side, and, with a little help, the patient gradually moves himself into the new bed.

Special precautions must be taken against the bed being

made wet. By underlaying waterproof materials (oil-skin or
india-rubber sheeting) this is much more easily done than
formerly, when such materials were not so common. But
they must be soft, must form no hard folds, and must be
often changed, washed, and aired. The patient should not
lie directly upon the waterproof material; it must be covered
with a multiple layer of linen—a linen sheet folded to a
suitable size as an underlay, and this, when wet, must be
changed. The change of the underlays must be made with
much consideration and care, so that no pain is given to
the patient. (This is usually done in the same way as described in the changing of the sheet.) The ends of the underlay (also called "draw-sheet") are tightly tucked under the
mattress, or they are firmly fastened with large safety pins,
to prevent the formation of creases. Smooth tension of the
underlay is best secured when the linen sheet is so doubled
that strong wooden rods can be slipped through the two
ends of the draw-sheet, between the folds (the two ends of
the underlay must be joined together like a round roller
towel). The rods then lie along the sides of the bed with
their ends projecting beyond the underlay at the top and
bottom. Now if a strong girth or strap be passed around
the upper ends of both staves, and another around the lower
ends of both, and these straps, running across under the bed,
are drawn tight (by means of buckles), then the formation
of creases in the underlay will be impossible.

We take for granted that the nurse always satisfies herself
that the bed-linen, mattresses, pillows, and coverlets are *aired
before* she lays them on the bed. In summer these articles do
not require warming—it is even pleasant to the patient to get
into a somewhat cool bed; but, in cold weather, the *bed
must be warmed* before he is placed in it: mattresses, pillows,

coverlets should be laid for some time in the sick-room, or in another warmed room. Sheets should be laid beforehand on a stove pipe, or close to the fire or stove until moderately warmed, but care must be taken that they are not scorched. If the patient is very weak and sensitive to cold, the undersheet, already spread over the mattress, must have hot bottles or bed-bricks repeatedly passed over it. If a particular part of the bed, the foot-end for instance, must be specially warmed, wrap a hot bottle in an underlay and place it there for a time, but do not forget to take it away before laying the patient in the bed, or he may burn his feet—a carelessness of which he may reasonably complain.

We shall now consider a condition that may become one of extreme pain to patients long confined to bed—may even cause death—*Bed-sores (Decubitus)*.

There are two very different species of Decubitus :

I. *The bed-sore* at the buttocks, at the shoulder-blades, and at the elbows. A good nurse has it almost wholly in her own power to prevent this ; the most liable are very thin persons, or those who frequently lie wet. Frequent cleansing, changing to dry clothes, and to another bed, in fact, all that we have previously described are the most important means for preventing bed-sores. Usually, even before the first symptom appears (a rose-pink spot the redness of which, by finger pressure, can be easily made to disappear, but returns immediately), the patient already complains of burning, and is frequently inclined to change his position. Twice daily at least, bathe the painful spots with cold water. If coldness be disagreeable, then use tepid water mixed half-and-half with brandy, or with vinegar. Moistening with fresh lemon juice is also excellent, and can be done in the simplest manner as follows : Cut a juicy lemon in two, and rub the cut surface over the red places.   Then lay a *soft* horse-hair pillow under the sensitive parts, or make the places under them hollow.   This is done with *seat* or *circular-centre cushions*, made of horse-hair covered with soft leather, or of india-rubber material inflated (air cushions); but these must not be too full, or they will be very hard for the patient.   He should never lie in direct contact with them.   They should always be covered by an underlay, but it must not be too thick.   When there are signs of bed-sores on the back, hips, elbows, or heels, the nurse must be able to make sufficiently large cushions, hollow in the middle, by sewing narrow bags, of requisite length and breadth, and filling them with horse-hair; failing this, they may be filled with wool, wadding, oakum, or jute.   Such a bag is formed into a ring by the ends being sewn together so as to make a hollow centre large enough to receive the painful part.

If, in spite of these preventive measures, excoriation appears, or if the bed-sores were formed before the nurse took charge of the patient, then bathing with lemon-juice, vinegar, or brandy must be omitted, because these produce pain at the sore parts.   The sores must be carefully bathed with lead-lotion, which agrees well with sores; the sore places must be dried by gentle pressure and dabbing with a soft, clean towel; then lay on a diachylon plaster.   Applying lead, zinc, and tannin (tannic acid) ointments is also good, but, as soft ointments must be spread on linen to keep them on the sores, and the small rags are easily rolled up and shift, plasters are preferable.   If the patient has lain himself sore at the base of his back, and can lie on his side part of the day, the healing is facilitated, especially if damp compresses of lead-lotion are applied every half hour or hour, but without wetting the bed.   With superficial excoriation, touching the sore spots with lunar caustic *(Lapis)* is of advantage. In favourable circumstances, this makes a thin, quick-drying, firmly-adhering scab, producing a cover over the wound, under which new epidermis is soon formed. Sometimes this touching will cause at first an intense smarting pain for some hours, but afterwards the patient will be compensated by rest and freedom from pain. The nurse must never make such cauterizations without special order from the doctor ; it is better that the caustic should be applied by the doctor, or the patient may think that the nurse has caused him pain from want of skill, or without necessity.

II. The second kind of bed-sore is the *gangrenous decubitus*. The nurse must *not* use this expression before the patient, because many of the unlearned, at the expression "gangrene," would at once seize the idea that they are irremediably lost. Here the term gangrenous only signifies that the skin looks as if burnt or charred. This form of "bed-sore" cannot always be prevented, as the cause lies in the diseases in which it principally occurs. In typhus fever, very severe inflammation of the lungs, small-pox, scarlet fever, also in spinal injuries, and in severe traumatic putrid fevers, when cleaning the almost unconscious or delirious patient, or when changing his bed, a more or less large, dark, blue-red spot in the region of the base of the back is discovered. This the nurse must at once bring to the doctor's notice on his next visit. The spot does not disappear when pressed with the finger ; it originated without pain, and increases, too, without causing pain. This decubitus proves that the blood runs feebly through the veins : the weight of the recumbent body on the part or parts affected suffices to bring the circulation to a standstill, and the blood stagnates, coagulates, and partly oozes into the skin. When such a blue spot is seen, care must be taken to prevent it from enlarging. The skin, saturated with blood in this manner and looking as if burnt, is always lost; when it falls off, an open sore is formed, and should the patient recover from his severe illness, this sore heals very slowly. In these serious cases I have found the water-cushion exceptionally serviceable. This is an india-rubber pillow with a tube fitted in on one side through which the water is poured; when sufficiently full it is closed, and the patient laid upon it. Of the utility and value of these water-pillows as aids to healing gangrenous bed-sores, very different opinions prevail among doctors. Some are greatly prepossessed in their favour—and I am one of them; others say most patients will not lie upon them because they are too uncomfortable, and they fear they will fall out of bed. They are costly, and that cost is increased from their being often torn, and cannot be satisfactorily repaired. To all these objections I can only say that the first cost is considerable (16 to 18 florins each), yet they are not wholly unattainable, or I should not have them in the hospital at my clinic. Their serviceable use depends on their proper management being understood, as follows :—

1. Choose only large water-pillows, which are almost bed-wide square.

2. Lay the pillow flat upon the middle of the bed, and fill it slowly with warm water (about 95° F. = 28° R. = 35° C.), through a funnel. Close the tube repeatedly, in order to test the tension of the pillow. This must be such that, by laying thereon both hands and arms, and pressing them down, effort is necessary to press the cushion together. Close the tube, and let it hang over the side of the bed.

3. Then spread an underlay over the water-cushion, and put pillows between the latter and the head of the bed, and another pillow at the foot-end (under the patient's knees).

F

4. Lay the patient carefully upon the water-pillow so that about a handbreadth of the lower end extends beyond the pelvis; then the sides and upper end of the cushion will swell, and the hollow of the loins be filled in.

5. Subsequent addition of warm water is unnecessary, as the heat of the body keeps the temperature of the water in the pillow sufficiently warm.

6. If the tension of the cushion is too slack, so that the patient presses it together in the middle, water must be added; if too tight, so that he lies hard, then, without lifting him, some water must be carefully let out. An empty water-cushion, covered with a sheet, can be laid under a patient and be filled whilst he lies upon it; this presents no difficulty, if the supply-tube be held high.

7. At each dressing, the sheet over the cushion must be changed.

8. If the cushion be no longer required, it is emptied by turning the tap, and then it is carefully lifted from the bed.

To lift or carry it, *never* take hold of the tube. To remove every drop of water, lay the cushion upon a sloping surface with the open tube at the lower end. Then place it in the cupboard,—where possible, unfolded and lying flat, or only rolled. Like all india-rubber materials, these cushions are liable to spoil if lying long in store unused; they then become brittle and break to pieces. When not likely to be used for a long time they should be filled at least once a month, and the water left in them for some hours.

If these instructions—the result of long experience—be followed, the patients will be satisfied with the water cushions, as they protect them from pains and perils much better than all circular-centre cushions. I am convinced my assistants and nurses have preserved the lives of many patients, suffering from gangrenous decubitus, by nursing them on water-pillows.

But this is not done by merely laying patients on the pillows. Around the patch of skin, becoming ever of darker blue until it becomes black, a bright red margin is formed in from eight to ten days; now at the edge of the quite dry and dead black skin humectation * begins, with sanious smell. The doctor must decide what dressing shall be used, or whether he will wait for the black skin to peel off of itself, or whether the dead shreds shall be cut away; this cutting will not be felt if the shreds be not pulled. If the convalescent has gradually gained strength, the dressing is easily done while he lies on his side; if too feeble, he must be held in the lateral position until the dressing is finished. On the black skin and dead shreds being loosened, the decubitus is said to be cleansed, and the suppurating surface is treated and dressed, like every other large wound, until it cicatrizes.

Our attention will now be given to that which is more gratifying : *the aid which a nurse can give to the convalescent patient, although still confined to his bed.* In a sick-room there must always be *a light* throughout the night. Gaslight in the sick-room I do not find injurious, although some doctors say they do. Formerly, when little was known of the management of gas-light, accidents occurred even in sick-rooms, but, with ordinary care, these cannot easily happen. Turning a gas-light low for the night does not diffuse a disagreeable smell; but the nurse must remember, if the light should be blown out by a draught of wind, at once to close the tap or the unburnt gas will escape into the room, and this may not

* " Humectation : the act or process of making moist."- MAYNE.

only be offensive, but may lead to suffocation or explosion. If the gas-light is placed so that it cannot be easily shaded, turn it off and replace it by some other kind of light; most patients are disturbed by seeing the light, it keeps them from falling asleep, or makes them wake easily. Petroleum lamps are useless for faint illumination at night; on turning them down a bad odour is emitted. Good oil-lamps are best, or floating night-lights, or the very thick stearine night-lights, with fine wick; but the uniform continuous burning of the last is so uncertain that their quality should be tested by experience before relying upon them. If the light goes out in the night and cannot be re-lighted, it is very troublesome. Tallow candles emit too strong an odour, and must not be used.

A non-striking clock of good size should be in every sick-room, as many patients desire to be able always to see it; it is much more simple and easy for the nurse punctually to carry out her orders with the clock before her, than if obliged constantly to look at her watch.

To make a patient comfortable when he is taking his food in bed he must be in as easy a position as possible. Above all, the nurse must remember that no one can drink with his head lying far back; it is painful to see fluids poured into the mouth in this position. If the patient must not be raised from the horizontal position, then, at least to drink, the head and the neck must be lifted a little. This is best done by the left arm being pushed under the nape of the neck, and, whilst so lifting and supporting him, the liquid is given by the right hand with a table-spoon, or with a small cup, but slowly and carefully, always waiting until he has swallowed. For the very weak the quantity of liquid passing from a table-spoon into the mouth is too much, especially if there be pain in the throat, and swallowing hurts them—then a tea-spoon should be used. If the patient can and may drink by himself, then his head and nape of the neck should be so supported by pillows that he can drink comfortably. For a recumbent patient the drinking vessel should never be more than half full, or some will be spilt. Patients who cannot raise the head, or who cannot or ought not to sit up, can drink through an india-rubber tube, one end being immersed in the liquid and the other placed in the mouth. For drinking, children's ordinary feeding-bottles are often more useful to such persons than glass or china boats.

When obliged for a lengthened period to take food in bed, from slow convalescence from severe illness, or from chronic disease, or from a broken leg, there must be no delay in making such arrangements as will enable the patient to take his meals with ease, and so to stimulate him to more comfort-able and more ample eating. Invalid dining-tables are made consisting of two boards or shelves each about 14 to 16 inches broad, and 32 to 36 inches long, joined at one end to a strong wooden pillar. The lower board (footboard) runs on castors so that the table can be easily moved in any direc-tion, or it can be pushed sideways against the bed, with the footboard under, so that the upper board comes over the top width of the bed and provides a firm table. The wooden pillar must be only two or three inches higher than the level on which the patient lies, and it must have an arrangement by which, on turning a handle, the table can be raised and brought into proper relation to the bed and the patient.

I cannot too strongly recommend these tables for every pri-vate house. Where they are not to be had, a firm wooden board should be supplied as long as the bed is wide, and about 20 inches in width. Fix supports at the narrow ends about

16 inches high, and place the table over the legs of the patient in bed. It must stand securely on the mattress, and be so arranged that slipping off sideways is not possible. Any carpenter can make it. Even if this little contrivance comes into use only for a week it will afford considerable relief to the patient, and will protect the bed from being soiled by food.

The nurse must not neglect to require patients to rinse the mouth after meals : particles of food remaining in a not too healthy mouth with a furred tongue, can, under some circumstances, become really injurious. She must also examine the bed after meals to remove bread-crumbs, &c., which may otherwise annoy the patient.

If so far recovered that he may *read in bed*, the nurse must insist upon his having good and suitable light. The light (daylight, gas, or lamp-light) must fall upon his book from behind, *over his head*. A small reading-desk can be placed upon the bed dining-table. If he sits in an arm chair, these various things must be attended to : one can scarcely conceive how much these little aids contribute to nurse his strength.

The nurse must always know how to set a patient up comfortably in bed. Generally, he is raised at the back a little, is supported by pillows, and so eats. If this arrangement makes him comfortable there is no objection to it, but it is really only *lying high*, not sitting. To place him in a sitting position, the patient is taken hold of below or by the pelvis, and then, with a quick, lifting movement, is pushed upwards in the bed ; thus, stooping forward a little, he *sits* in the bed, the pelvis is upright, the hips are bent, and he is no longer recumbent. If strong enough, he can sit up of himself, but many persons require much attention before getting into the sitting posture. When bandaging the upper part of the body of the bedridden patient it must never be forgotten that, where possible, he must be brought out of the lying high position into the sitting posture, as otherwise, in spite of back-support (very objectionable in bandaging), he will always exert himself much more than when in the proper sitting posture.

To *lift a recumbent patient upwards* in the bed who has slipped towards its foot, by grasping him under the arms, is very awkward ; it is very difficult for the nurse to lift in this way, as then the patient is only " tugged," and little good results. By sliding her arms under the pelvis she can easily lift him, especially if he assists by pressing with his hands and heels against the bed.

Female nurses' schools, and the religious orders that engage in sick-nursing, are reproached with sending assistance only to hospitals and into the houses of moderately well-to-do families, and with really doing nothing for the *sick-nursing of the poor* in their homes. At the first glance this reproach seems just. But if those who raise it will reflect a little they will soon be convinced that their demand, at least for the present, is impracticable. It is evident that the essential conditions of rational and successful sick-nursing, such as good air, light, warmth, bedding, good food, &c., are altogether wanting in the homes of the poor ; in fact, they are not to be found even in most of the dwellings of the lower middle, and mechanic classes.

Of what use is the gratuitous supply and regular giving of medicines if every necessary is wanting even for ordinary healthy living? It is not that the nurse and the nun shrink from enduring the privations and injurious influences existing in the cottages and hovels of the poor, as well as the coarseness of the lower classes, but it is the impossibility of being useful, under such circumstances, that renders home-nursing unattainable for the poor. For this reason hospitals of various descriptions exist for the reception of the poor and less well-to-do. In those classes, too, in which, although the earnings suffice to maintain even a numerous working family household when in a state of health, yet everything is unhinged when a member falls ill: the shrinking from admittance to the hospital must disappear. Certainly, to many it is very hard to be severed from husband or wife, brothers or sisters, children or parents for a time, but, in addition to this there is the fear of encountering coarse, rough, avaricious hospital nurses, nay, it is the very tone prevailing in many hospitals, where the sufferers are regarded and treated as "cases," as ciphers, that keeps many away from hospital treatment.

The time must come when hospitals will be so arranged, and patients so well treated, that those entering will not do so from sheer necessity, but from the conviction that they will recover much more quickly there than if the doctor occasionally visited them in their own homes, where, at times seeing no possibility of rendering them any efficient help, in their circumstances, he either leaves them to themselves, or, to pacify his own conscience, he prescribes a number of remedial measures that the relatives of the patient are absolutely unable to carry out. One can comfort the poor in their cottages, and give them food and medicine gratuitously, yet to nurse and treat them there with any prospect of success cannot be done unless wealthy associations are formed so that they can, at once, provide a healthy dwelling, good food, &c., for the whole family—in short, can change the poor into the condition of well-to-do people, at least during the illness. This would be splendid if it were possible, but at present we must be content with more modest limits to our humane efforts. We must establish more hospitals, and strive to make them more comfortable, with still better female nurses, so that the repugnance of the middle classes to enter into them may be overcome.

Observations and Excellent Remarks upon the Peculiarities of many Patients, and upon the Nurse's Consequent Conduct,

BY FLORENCE NIGHTINGALE.

The book * of this celebrated English Sick-nurse contains many admirable observations and remarks upon the peculiarities of many patients, and upon the nurses. I could not better express and combine them than she has done, and therefore I give the following extracts in the words of this excellent lady, who has done so much for the improvement of sick-nursing :—

"It is as impossible in a book to teach a person in charge of sick how to *manage*, as it is to teach her how to nurse. Circumstances must vary with each different case.

"How few men, or even women, understand, either in great or in little things, what the being 'in charge' is. From the most colossal calamities down to the most trifling accidents, results are often traced (or rather *not* traced) to such want of some one 'in charge,' or of his knowing how to be 'in charge.'

"To be 'in charge' is certainly not only to carry out the proper measures yourself, but to see that every one else does so too; to see that no one either wilfully or ignorantly thwarts or prevents such measures.

"Unnecessary noise, or noise that creates an expectation in the mind, is that which hurts a patient.

"Never to allow a patient to be waked, intentionally or accidentally, is

---

* "Notes on Nursing," by Florence Nightingale. London : Harrison & Sons.

a *sine quâ non* of all good nursing. If he is roused out of his first sleep, he is almost certain to have no more sleep. It is a curious but quite intelligible fact that, if a patient is waked after a few hours' instead of a few minutes' sleep, he is much more likely to sleep again.

"A healthy person who allows himself to sleep during the day will lose his sleep at night. But it is exactly the reverse with the sick generally ; the more they sleep the better will they be able to sleep.

"A whispered conversation in the sick-room is absolutely cruel ; for it is impossible that the patient's attention should not be involuntarily strained to hear.

"Affectation, like whispering or walking on tip-toe, is peculiarly painful to the sick. An affectedly quiet voice, an affectedly sympathising voice, sets all their nerves on edge. Better almost make your natural noise.

"Some nurses cannot open the door without making everything rattle. Or they open the door unnecessarily often, for want of remembering all the articles that might be brought in at once. I have seen an expression of real terror pass across a patient's face whenever a nurse came into the room who stumbled over the fire-irons, &c.

"A good nurse will always make sure that no door or window in her patient's room shall rattle or creak ; that no blind or curtain be made to flap—especially will she be careful of all this before she leaves her patient for the night.

"Always sit within the patient's view, so that when you speak to him he has not painfully to turn his head round in order to look at you. Everybody involuntarily looks at the person speaking . . . ; so, by continuing to stand, you make him continuously raise his eyes to see you. Be as motionless as possible, and never gesticulate in speaking to the sick.

"Never speak to a sick person suddenly ; but, at the same time, do not keep his expectation on the tip-toe.

"Do not meet or overtake a patient who is moving about in order to speak to him, or to give him any message or letter. You do not know the effort it is to a patient to remain standing, for even a quarter of a minute, to listen to you.

"Everything you do in a patient's room, after he is 'put up' for the night, increases tenfold the risk of his having a bad night. But, if you rouse him up after he has fallen asleep, you do not risk, you secure him a bad night.

" Remember never to lean against, sit upon, or unnecessarily shake, or even touch the bed in which a patient lies. This is invariably a painful annoyance.

"Conciseness and decision are above all things necessary with the sick. Let what you say to them be concisely and decidedly expressed. What doubt and hesitation there may be in your own mind must never be communicated to theirs, not even (I would rather say, especially not) in little things. Let your doubt be to yourself, your decision to them. People who think outside their heads, who tell everything that led them towards this conclusion and away from that, ought never to be with the sick.

" Irresolution is what all patients most dread. Rather than meet this in others, they will collect all their data, and make up their minds for themselves.

"Above all, leave the sick-room quickly and come into it quickly, not suddenly, not with a rush. But don't let the patient be wearily waiting for when you will be out of the room or when you will be in it. Conciseness and decision in your movements, as well as your words, are necessary in the sick-room, as necessary as absence of hurry and bustle. To possess yourself entirely will ensure you from either failing—either loitering or hurrying.

" With regard to reading aloud in the sick-room, my experience is, that when the sick are too ill to read to themselves they can seldom bear to be read to.

"If there is some matter which *must* be read to a sick person do it slowly. People often think that the way to get it over with least fatigue to him, is to get it over in least time. They gabble ; they plunge and gallop through the reading. There never was a greater mistake.

" If the reader lets his own attention wander, and then stops to read up to himself, or finds he has read the wrong bit, then it is all over with the poor patient's chance of not suffering.

" To any but an old nurse, or an old invalid, the degree would be quite inconceivable to which the nerves of the sick suffer from seeing the same walls, the same ceiling, the same surroundings, during a long confinement to one or two rooms.

" I shall never forget the rapture of fever-patients over a bunch of bright-coloured flowers. I remember (in my own case) a nosegay of wild flowers being sent me, and from that moment recovery becoming more rapid.

" It is a matter of painful wonder to the sick themselves, how much more they think of painful things than of pleasant ones ; they reason with themselves, they think themselves ungrateful ; it is all of no use.

" A patient can just as much move his leg when it is broken as change his thoughts when no help from variety is given him. This is, indeed, one of the main sufferings of sickness ; just as the fixed posture is one of the main sufferings of the broken limb.

" We will suppose the diet of the sick to be cared for. Then, this state of nerves is most frequently to be relieved by care in affording them a pleasant view, a variety of flowers, and pretty things. Light by itself will often relieve it. The craving for ' the return of day,' which the sick so constantly show, is generally nothing but the desire for light, for the relief which a variety of objects before the eye affords to the harassed sick mind.

" Again, every man and every woman has some amount of work with the hands, excepting a few fine ladies, who do not even dress themselves, and who are really, as to nerves, very like the sick. Now, you can have no idea of the relief which such manual labour is to you—of the degree to which the being without it increases the peculiar irritability from which many invalids suffer.

" A little needlework, a little writing, a little cleaning, would be the greatest relief the sick could have, if they could do it. Reading, though it is often the only thing the sick can do, is not this relief. Bear also in mind to obtain for them all the varieties which they can enjoy.

" I need hardly say, that too much needlework, or writing, or any other continued employment, will produce the same irritability that too little produces in the sick.

" With most very weak patients it is quite impossible to take any solid food before 11 A.M., nor then, if their strength is still further exhausted by fasting till that hour.

" A spoonful of beef-tea, of arrowroot and wine, of egg-flip, every hour, will give them the requisite nourishment, and prevent them from being too much exhausted to take, at a later hour, the solid food which is necessary for their recovery.

" Again, a nurse is ordered to give a patient a teacupful of some article of food every three hours. The patient's stomach rejects it. If so, try a tablespoonful every hour ; if this will not do, a teaspoonful every quarter of an hour.

" In very weak patients there is often a nervous difficulty of swallowing, which is so much increased by any other call upon their strength that, unless they have their food punctually at the minute, which minute again must be arranged so as to fall in with no other minute's occupation, they can take nothing till the next respite occurs—so that, an unpunctuality or delay of ten minutes, may very well turn out to be one of two or three hours.

"But, in chronic cases, the consulting the hours when the patient can take food, the observation of the times, often varying, when he is most faint, the altering seasons of taking food in order to prevent such times—all this, which requires observation, ingenuity, and perseverance (and these really constitute the good nurse), might save more lives than we wot of.

"Exhaustion from a half-starvation is one of the most frequent causes of loss of sleep. Many a patient will sleep exactly in proportion as he can eat.

"To leave the patient's untasted food by his side, from meal to meal, in hopes that he will eat it in the interval, is simply to prevent him from taking any food at all. Patients have been literally made incapable of taking one article of food after another by this piece of ignorance. Let the food come at the right time, and be taken away, eaten or uneaten, at the right time ; but never let a patient have 'something always standing' by him, if you don't wish to disgust him of everything. *

"That the more alone an invalid can be when taking food the better, is unquestionable ; and, even if he must be fed, the nurse should not allow him to talk, or talk to him, especially about food, while eating.

"One very small caution,—take care not to spill into your patient's saucer,—in other words, take care that the outside bottom rim of his cup is quite dry and clean ; if, every time he lift his cup to his lips he has to carry the saucer with it, or else to drop the food upon and to soil his sheet, or his bed-gown, or pillow, or, if he is sitting up, his dress, you have no idea what a difference this small want of care on your part makes to his comfort, and even to his willingness for food."

---

* A lady once told me that a sick-nurse had made the taking of soup quite impossible to her, because the nurse would always taste it in her presence. This the nurse must never do. The food should always be brought into the room so that the patient can at once take it ; tasting and cooling must be done outside the room. Should it be, for once, exceptionally unavoidable to taste it in the presence of the patient, then let the nurse take a clean spoon ; but, having once tasted soup therewith, it must not be again placed in the plate, because even such indirect contact with any one is objectionable to many persons.—Dr Th. B.

CHAPTER III.

ON THE FULFILMENT OF MEDICAL ORDERS.

NOTHING is more wearing to the doctor than to be compelled repeatedly to explain in detail how his orders should be fulfilled. With the utmost patience he may repeat twenty times a day how the prescribed powders shall be taken, or the drops instilled into the eye, or the compress applied, &c.; yet his hearers, agitated by anxiety and affliction, listen with half an ear only, so that often he finds his thorough explanation not understood in the least, or misunderstood. Although it is painful to have to repeat everything again at the next visit, it is far worse to find that, in the meanwhile, everything necessary for the patient had not been done, or something prejudicial had been done instead.

When no special directions have been given in particular cases, the following Instructions *On the Fulfilment of Medical Orders* are for the guidance of the nursing staff :—

1. ADMINISTERING MEDICINES.

It is not caprice on the part of the doctor when he

prescribes the medicines in solutions, in powders, or in pills, but it depends partly on the quantity, partly on the nature, of the substances to be incorporated by the body as to the form in which they may be most judiciously conveyed so as to secure the desired effect. We refer, in the first place, to such medicines as pass into the stomach through the mouth, to operate either directly upon the mucous membrane of the stomach and intestines, or to be thence absorbed into the blood, and, through the blood, to operate upon other organs. With *liquid medicines* it is ordered, as a rule, that they shall be administered at definite periods in doses of a tablespoonful, a teaspoonful, or a stated number of drops. These measures are somewhat indefinite, on account of the variation in size of the spoons. Usually a tablespoonful is 4 fluid drachms (about 16 grammes); a teaspoonful, 1 fluid drachm (about 4 grammes). One gramme weight of water (nearly $15\frac{1}{2}$ grains) is equal to the measure of a cubic centimetre of water. If most medicines are somewhat heavier than water, yet the single dose is so calculated that this small difference is not injurious. It is very desirable that the method of measuring single doses in small measuring glasses should be more generally introduced, at least into hospitals, and, by means of the nurses, into families also. The glasses used should contain about 4 fluid ounces; the lines and figures engraved thereon should mark clearly the measure of the contents in gradations of teaspoonfuls. Spoons have this disadvantage that, when full, they must be held with a very sure hand, or the medicine will be spilt over the patients or the bed linen. Some patients, accustomed to taking medicine from spoons, may object to the measure-glass. For them, small and large china spoons should be provided, the one to hold somewhat more than a teaspoonful, the other somewhat more than a

small tablespoonful, and thus neither will need to be quite
filled, and spilling the medicine will be prevented. Giving
medicine *in drops* requires particular care and a sure hand.
Let the drops fall upon a piece of sugar, or into water in a
teaspoon or a glass. To give the drops equal in size and of
the exact number prescribed, special *minim-glasses* are made.
They are finely-pointed glass tubes into which the liquid is
drawn ; by pressure upon an india-rubber cap strained over
the upper opening of the glass, as many equal-sized drops
can be expelled as are required.

At times the *preparation of an infusion* is left to the patient
himself, or to those about him ; the doctor notes only,
"for an infusion." Then, as a rule, a heaped teaspoonful is
taken to a cupful of boiling water. It is made in a teapot
and left to stand for a quarter of an hour on a warm stove or
fireplace, without boiling ; the water is strained off and drunk.
If more than one cupful is prescribed, an adequate additional quantity is taken of the herbs, seeds, flowers, or cut-up
roots ordered by the doctor. Besides camomile and elder
flowers, those most frequently prescribed for infusions are
fennel, valerian, marsh-mallow *(Althæa officinalis)*, walnut
leaves, pansy, buck-bean, and senna leaves.

When not detrimental to the proper effect of the medicines,
the doctor sometimes adds fruit-syrups, aromatic waters, &c.,
to improve the flavour. He cannot always ask the patient
which he prefers, and it may happen that the flavour of the
medicine is objected to. When this is the case, and, from
aversion, the patient vomits the medicine, the nurse must tell
the doctor. Medicines of unpleasant flavour are often more
easily taken when mixed with some selters or soda water.
Persuasion, as with children, often helps. Unpleasant after-
taste of medicine is best removed by drinking fresh water, or

by rinsing the mouth with fresh water to which a little cognac or essence of peppermint has been added. If the scent of a medicine be repugnant to the patient, he must keep his nose closed whilst taking it.

Many patients (adults more than children) strongly object to taking oily substances, such as castor oil or cod-liver oil. Various means have been devised to make the taking of these oils less objectionable. For instance, castor oil may be mixed in a cupful of hot beef-tea, or the patient may be induced to chew some roasted coffee-beans, or to suck a slice of fresh lemon before and after swallowing the oil. To many, cod-liver oil becomes less unpleasant by adding a little ale or black coffee. Oily medicines can now be had enclosed in gelatine capsules. Acids, and many ferruginous mixtures, are said to affect the teeth injuriously. If these be taken for a short time only, it is not of much consequence. I should like to believe that it is nothing but the sensation of the teeth being set on edge that gives rise to this opinion. However, advise patients anxious on this point to place the spoon with the acid liquid directly at the back of the tongue, so as not to touch the teeth, and at once to rinse the mouth with water; or the doctor may be asked whether the prescribed dose may not be put into a glass partly filled with water, and taken diluted. Fluids, the contact of which with the teeth is feared, may be drawn up through glass tubes; or small gelatine capsules, formed like a box and filled from the minim-glass, may be swallowed like pills. Further, the doctor must be asked if the remedies are to be taken on an empty, or on a partially-filled, stomach. With substances which, even diluted, slightly excite the mucous membrane of the stomach, such as arsenic drops, the latter is decidedly preferable. In acute, severe illnesses, it is seldom necessary to disturb the sleeping patient in the night to give the prescribed medicine, but on this point the doctor must be consulted.

Powders are prescribed to be given in various ways. When the *exact* quantity of the powder is not material, the prescription states, for instance, thus : " a teaspoonful," or " a heaped teaspoonful to be taken." In the first case, a teaspoon is filled with the powder, and levelled off by the back of a knife to the edges of the spoon ; in the second case, the powder is heaped up in the teaspoon. Most powders thus prescribed are taken dissolved, or in suspension, in a glass full, or half full, of water. Fully prepared *effervescible powder* (in which the ingredients are already mixed by the chemist) must be kept in a glass bottle with a large mouth having a glass stopper ; this must be set in a dry place or it will absorb the air-moisture, and on being dissolved will not effervesce. It is necessary to do the same with most powders containing sugar, specially with hygroscopic * remedies, such as the bromide of potassium. On giving an effervescible powder, take an ordinary drinking-glass and fill one-third of it with water ; then, with a *dry* spoon, put a heaped-up teaspoonful of powder from the glass bottle quickly into the water, stir quickly, and let the patient drink it whilst effervescent. It is the gas thereby evolved (carbonic acid) that the patient is to swallow with the liquid.

In spite of all care, effervescible powders are liable to absorb moisture if compounded some time before use ; they are often prescribed so that one part (tartaric acid) is in one paper, and the other part (bicarbonate of soda) is in another ; the two combined contain enough for one dose in a glass of water ; usually one powder is in white paper and the other in

---

* Hygroscopic remedies, *i.e.*, remedies having "the property of attracting or giving off moisture."—MAYNE.

blue. The glass of water being ready (sugar may be added to improve the flavour) put in the bicarbonate of soda, and when dissolved, add the tartaric acid; whilst effervescing, stir and drink. Or the two powders may be separately dissolved in water in different glasses; when dissolved, pour from one into the other and drink whilst effervescing.

The prescription, that a *knife-pointful* of powder is to be taken, is clear in itself but the measure is inexact, and the doctor must be asked whether it is to be heaped or otherwise.

When much depends upon the accuracy of the quantity of the medicaments to be taken in each dose, the powders are weighed and divided into separate packets by the chemist.

To take a powder, the simplest way is to let it fall direct upon the tongue, and to swallow it by drinking water afterwards. But many patients cannot do this, and it must not be so given to patients seriously ill, or to children; for them it must be dissolved in water, or, if not soluble, must be mixed with water in a table-spoon. A glass rod, or the handle of a teaspoon may be used for mixing,—the nurse must never use her finger, because that is nauseating to every patient.

Powders of unpleasant flavour (quinine, for example) are often enclosed in *wafers*.*   Cut a piece of wafer from 2 to $2\frac{1}{2}$ inches square, put it into a tablespoon, moisten with a few drops of water, and shake the powder on to its centre; then fold the wafer over the powder, let the patient lay it upon the back part of his tongue, and, by drinking water, it is easily swallowed. Chemists can supply powders in wafer capsules.

Many patients are very awkward in *taking pills*—they should place the pills quite at the back of the tongue and swallow

* The Wafers are obtainable from most chemists and from many confectioners, but in the country, chemists and confectioners are not always at hand. In case of need, *very fine rice paper*, well moistened, may be used as a substitute.

them with water. Children seldom swallow pills without biting them.

It is scarcely necessary to lay special stress upon this point, that *vessels used in giving medicines must be perfectly clean, and, immediately after use, must again be most carefully cleansed.*

It is important to refer to the *assistance that may be given during vomiting* after emetics. Emetics are generally prescribed in the form of powders or liquids, with the direction, "till copious vomiting results." If the patient vomits very quickly after the first dose, and ejects a great part of the emetic, then a second dose must soon be given. If the stomach be soon emptied by the vomiting, and the retching continues, drinking camomile tea will relieve the patient. In the act of vomiting it is agreeable to most patients for the nurse to lay her hands on the front and back of the head to support it. If the vomiting continues longer than desired, give five or six drops of tincture of opium on a piece of sugar, or in water, and lay a mustard plaster on the pit of the stomach.

Purgatives, which operate quickly, as senna (Viennese black-draught), mineral waters, and enemata, are given in the morning or at mid-day, so as not to disturb rest at night. Laxatives, which operate slowly, such as small doses of rhubarb, aloes, or castor-oil, may be given in the evening, as their action begins the next morning, or later in the forenoon.

2. INHALATIONS.

The inhaling of medicated substances has recently come into more general use. To inhale very volatile substances, as ammonia (spirits of hartshorn), acetic ether, eau de Cologne, &c., no special apparatus is necessary: a little bottle filled

with one of these is held to the nose, or it is put upon a handkerchief and thence inhaled. Slightly volatile, or non-volatile,
aqueous solutions of medicaments are dispersed by special
apparatus (spray-producer, vaporizer, hand-spray, steam spray,
&c.) in the form of fine drizzling rain, which must be directed
against the open mouth or nose of the patient. The nurse
must be entrusted with the arrangement of the apparatus, so
that she can not only prepare it for use, but she must fully
understand how to discover quickly the cause of its not acting
properly.

In diseases of the respiratory organs the small steam spray
apparatus is mostly used; the hand-spray is useful for giving
refreshing coolness in the sick-room by spraying with it cold
water, or alcoholic aromatic waters.

In the absence of special apparatus (inhaler), the *oil of
turpentine* may be inhaled in the following way: put hot
water into a small vessel, and add some drops of oil of
turpentine; fix a strong paper screw bag by its open end
over the vessel, and cut away a piece of the point at the small
screw-end of the bag—to this opening, place the mouth or
nose of the patient, so that, by deep inspirations, he can inhale
the oil of turpentine mingled with the steam. To effect the
same object, respirators are made. These are small appliances
fixed before the mouth, having cotton wool fitted between two
sieves, through which the air to be inhaled must pass. The
volatile substance for inhalation is dropped upon the cotton
wool. When inhaling from vessels containing hot water,
children or weak patients must never be left alone, for the
vessels have been upset, and bad scalds have resulted. The
apparatus must be so placed as to require no exertion on the
part of the patient; how long he is to inhale at a time, and
how often, must be determined by the doctor.

3. INJECTIONS; ENEMATA; SUPPOSITORIES; INSTILLATIONS.

To purify cavities of the body, or to introduce into them medicaments dissolved in water, the *syringe*, and so-called *douche-apparatus*, are used. To dispense with the use of syringes is desirable, because much care is requisite to keep them in order. When constantly used they mostly act well, if the piston is occasionally oiled : if seldom used, the piston dries ; this must then be drawn out of the tube, and laid in water until swollen sufficiently to fit air-tight. For injections into the ear and nose, the ball-syringe is useful. The india-rubber ball, to which a thin ivory tube is fixed, is compressed, the open end of the tube is put into the liquid, and, on relaxing the pressure, the liquid is drawn in. In so doing the ball seldom gets quite full—it always contains a little air. Before injecting, compress the ball to expel the air, and then inject the liquid into the cavity by continuing the pressure. Practice is necessary for this, so that, by equal pressure more or less strong, the force with which the liquid is to flow and to penetrate the cavities of the body shall be completely governed. In selecting syringes the nozzle must be carefully examined to see that it is smoothly rounded off, and is without sharp or pointed edges, or slight injuries may thence arise.

For cases requiring injection of a larger quantity of liquid, small pumps are made. These are placed in a hand-basin of water, and the piston is managed by the patient, or they are filled by the turning of a crank, and, with moderate pressure, empty automatically (Clyso-pump).

Of late, efforts have been made to replace syringes by *douche-apparatus* (also termed *irrigators*), allowing of more or less high pressure. These irrigators are simple in construction, and are made as follows : a cylindrical vessel, about 10 inches high and 4 inches in diameter, of japanned sheet-iron (best if enamelled on the inside), or of glass, with removable cover, has a short pipe leading from the foot of the vessel, and on to this is drawn a suitably thick indiarubber tube, about 40 inches long, having a syringe-nozzle of ebonite inserted at its extreme end, closed by a stop-cock. A strong ring at the upper edge of the vessel is convenient by which to suspend it when necessary.

For example, when using this apparatus for syringing the nose (nose-douche), suspend it from the wall over the wash-stand at such a height that the patient, standing with his head bent forward (so that the water runs into the hand-basin), can easily convey the special syringe-nozzle at the end of the pipe into his nostril ; when the stop-cock is opened, the water spurts into the nose with moderate force. By opening the stop-cock more or less, or by hanging the reservoir lower or higher, the pressure can be regulated at will. The same apparatus may be used for an eye-douche by affixing a differently formed syringe-nozzle, perforated like a sieve. A douche-apparatus can also be made by taking a basin, a jug, or a bottle, if the pipe with the stop-cock at one end has its other end so conducted over the edge of the vessel that it lies at the bottom of the vessel in the liquid. Care must be taken that the tube does not kink where it passes over the edge of the vessel, and this is prevented by inserting a short, stiff knee-joint in the tube at that point.

When filled, the tube acts as a suction-pipe, and the liquid streams out so long as the end in the vessel is immersed, and so long as the other end, with open stop-cock, is held lower than the surface of the liquid in the vessel.

It is important to consult the doctor whether the liquid

to be injected must be very cold or very warm, or cool or tepid; if the pressure must be strong, weak, or only moderate, for the effect obtained is very different from the various gradations of temperature and of pressure of the fluid upon the interior of the nose, the eye, or the ear.   More especially, too cold or too warm injections into the ear must be guarded against; these must never be given with strong pressure by the inexperienced.

Special caution is necessary when *injecting into the rectum.* This is termed, "to give a lavement," or, "to give an enema." No one should attempt it who has not already seen the little operation performed by an experienced person.   Wounds, even complete perforation of the rectum, often result, from which the patient may die.   He does not easily wound himself when introducing the nozzle affixed to the pipe of a clysopump, and then pumping in the fluid, or, after the piston has been drawn up, letting it be driven in by the pump itself. Pliable enema-nozzles are made, and may be confidently introduced.   Those of hard material must have a bulbous end so as not to injure.   The little bone tubes supplied with pewter syringes are too thin and pointed, and are dangerous.

Injections into the rectum have various objects; most frequently to expand the intestine with the fluid, and bring it into movement so that it shall expel, not only the injected fluid, but the excrement at the same time.   These are termed *laxative enemata.*   Warm water of about 95° F. ($=28°$ R. $= 35°$ C.) to which some olive oil is added (1 or 2 tablespoonfuls), is generally used.   The quantity injected should not exceed from 7 to $10\frac{1}{2}$ fluid ounces for adults (for children $1\frac{3}{4}$ fluid ounces; babies $6\frac{1}{2}$ fluid drachms), or it is too quickly expelled; it must be injected slowly.   When

ordinary enemata have been inoperative, it is all the more
necessary not to inject too much fluid at once. Substances
are then added (according to medical instructions: castor-
oil, infusion of senna leaves, honey, salt, soap), which, for
a while, are to exert an influence upon the inactive intestine.
Should these remedies be unsuccessful, then more injections
of larger quantities of water are sometimes used, so that
the water shall not only remain in the rectum, but shall
be driven higher up into the intestine. This is not done
unless specially ordered, and is attended with certain
success only when the doctor succeeds in inserting a longer
flexible tube, with olive-shaped head, high up into the
intestine of the patient, who must lie on his side or in the
knee-elbow position. To the projecting end of the tube,
when no irrigator is at hand, an indiarubber pipe is fixed,
about 40 inches long; this is held up perpendicularly, with a
large funnel inserted into its upper end. A person standing
on a chair slowly pours the water into the funnel. As the
water sometimes rushes out with great force by the side of
the pipe, but always after its removal, the patient must previously be so placed that it can pass direct into a large vessel
underneath (pail, tub, or shallow vessel); the bed must be
protected against saturation by large underlays of waterproof
material. As the patient may be greatly cooled by this
annoying, if not painful, operation, provision must be made
for quickly warming him again.

Starch enemata require special notice. With the addition
of some solution of nitrate of silver, as medically prescribed,
these are often successfully used, particularly with children
suffering from catarrh of the rectum. A thin paste of starch
is prepared, the prescribed quantity of nitrate of silver is then
added, and the fluid allowed to cool until lukewarm, then it

is drawn into the syringe (which must be of glass or of india-rubber), and about $1\frac{3}{4}$ oz. to 2 ozs. injected. (For adults, double the quantity.)

The number of remedies that may be introduced into the rectum by the syringe is not large; they are solutions of astringent substances, as alum, tannin, and acetate of lead; to allay spasm and pain, valerian and camomile infusions, opium, belladonna, chloral, but none of these must be used unless specially prescribed by the doctor.

For patients who can retain neither food nor fluid in the stomach, and for those who, suffering from contractions or from complete constriction of the œsophagus, are unable to convey anything into the stomach, nourishing substances are introduced through the rectum. These *nutritive injections*, which are to be absorbed from the rectum, must be small in quantity each time ($3\frac{1}{2}$ fluid ounces), or they will be expelled. For this purpose, milk, strong bouillon, malt extract, or wine, is used. Recently, artificially half-digested albumen and meat (solutions of peptone) have been given; but the nourishing effect of these injections is certainly inconsiderable, and many patients cannot long continue their use, because the rectum gets into a painful catarrhal condition; yet by these injections the lives of patients may be sometimes protracted until further assistance is obtained.

Suppositories are now seldom used, and when used, mostly for children. Cut a piece of firm ordinary washing-soap from 1 to $1\frac{1}{2}$ inches long, as thick as a stout lead pencil, and thin it off at one end; as it is, or, after being dipped in oil, insert it slowly into the anus. This excites a slight mechanical stimulus upon the walls of the intestine, and stimulates it to contractions, to evacuations of excrement, with which the soap is expelled. Sometimes suppositories are used to enable medicaments (morphia, belladonna, tannin, &c.) to act upon the abdominal organs, and to be incorporated into the body generally. Those prepared by chemists consist mainly of cocoa-butter; the thin end is dipped in oil, inserted into the anus, where it melts, and the cocoa-butter, with the medicament, is absorbed by the mucous membrane of the intestine.

Into the eye (more properly speaking, into the conjunctival sac), and into the ear (into the external auditory canal), *instillations* of medicinal solutions have often to be made. For an *instillation into the eye*, the patient lies upon his back, and the nurse lets the liquid drop into the opened eye from a small bottle or minim-glass, but in the region of the inner corner of the eye. Should he involuntarily close his eye, the liquid remains standing as a little lake in its inner corner; in this case the patient must remain lying, and again open his eye, when the fluid will enter. A simpler way is, to hold a small camel's-hair brush, plentifully saturated with the eye-wash, with its point to the exposed edge of the slightly drawn-down lower eyelid: at the moment of contact the drop flows in. As eye-washes are mostly weak solutions of medicaments, it is not very important whether more or less enters the conjunctival sac; the patient, therefore, may himself dip his clean forefinger into the fluid, and, in a recumbent position, repeatedly moisten the eye, opening and shutting it so that the eye-wash enters the conjunctival sac. For *instillations into the ear* the patient must lie upon the side that is well, and so remain for a time; after the instillation, cotton-wool must be put into the ear, at least for a short time, so that the fluid instilled does not run out and down the neck.

4. COMPRESSES. APPLICATIONS WITH A CAMEL'S HAIR PENCIL. EMBROCATIONS. MASSAGE. GYMNASTICS. ELECTRICITY.

Damp compresses are ordered for very different purposes—often only to warm, or to cool, certain parts of the body. We shall refer to this in the chapter on the Continuous Use of Cold and of Heat. For inflamed parts or for wounds, most frequently compresses of *lead-lotion* are ordered. With all compresses special care must be taken that they do not become dry, and that the linen, clothes, and bed are not made wet. Fold linen or cotton material in eight layers one over the other, and so large that it overlaps the part to be covered by about 1½ inches on all sides. The very loose cotton material of which bandages are generally made (bandage-gauze, mull, calico) can also be used, but then it must be folded in 12 layers one over the other. The English bandaging material (a coarse, loose cotton material similar to flannel) in two or three layers, one over the other, can likewise be used.

These layers of material *(the compress)* are dipped into, or wetted with, the lotion ; they must be fully saturated, but when to be laid upon sound skin, and to remain some time, they must be wrung out so as not to drip. A piece of waterproof material (oiled silk, or the finest indiarubber tissue, or gutta-percha paper) is placed over the compress, so as to overlap it about an inch and a half on all sides. Fasten this with a dry handkerchief, or a bandage, so that it shall not move from its place, and such a compress must be renewed every two hours, unless the doctor orders otherwise.

If compresses are to be applied to wounds or sores, they must not be wrung out ; and if the part affected can lie exposed and rests, the compress need not be fixed. The

waterproof covering is laid lightly over the compress, which
must be renewed every quarter or half hour, if not contrary
to the doctor's order.   If frequent renewal is unnecessary, and
the patient can walk about with the dressing on, it must be
fastened as described.   In these cases the compresses become
soiled with pus, and two at least are requisite—one on the
wound, the other lying ready in the lotion ; that taken from
the wound must be carefully washed in clean water before it
is again placed in the lotion.

Applications with Camel's Hair Pencils are prescribed :—

A. On the uninjured skin.   Tincture of Iodine is most
frequently used, diluted, pure, or concentrated, according to
the effects at which the doctor aims.   The skin of different
persons is sensitive to this remedy in very different degrees,
so that it is not always possible to foretell the actual effect that
will result ; in general, children and fair women are more sensitive to it than older persons and dark-complexioned men.   As
a rule, paintings with tincture of iodine are made to effect
the dispersion of chronic external swellings.   The remedy
should be continued for some time, and, consequently, must
not be permitted to cause either vigorous inflammation of the
skin, or pain.   In practice I have found it better to mix the
tincture of iodine with an equal quantity of tincture of galls,
and to apply the same in periods of three days, with an interval
of three days.   The painting to be done *once* daily (in the
evening), on three successive evenings, then a pause of three
days, then again paint for three evenings, and so on for weeks
and months.   The part must be painted with the tincture
each time until it is a deep dark-red brown.   When the
painted part permits, cover it with a piece of finest india-
rubber tissue and fasten with a bandage.   This is to prevent the iodine from being immediately rubbed off before

it has completely penetrated the skin, and to keep the linen from being stained blue. (By washing with soap the blue stains disappear; if the fingers be stained brown by iodine, liquid ammonia—*solution of ammonia*—will remove the stain.) Occasionally, wet dressings of the diseased part are ordered with the paintings with tincture of iodine; first cover the painted part with waterproof material, then apply wet dressings, and over these place a large piece of indiarubber material, and fasten the whole dressing with a large handkerchief or bandage.

By this method of using the tincture of iodine the deep brown-stained epidermis gradually becomes harder, leather-like, shiny, and then cracks. On the tincture penetrating through the cracks into the softer layers of the epidermis, intense smarting sometimes ensues; in this case, wait some days until the dry epidermis may be removed in shreds; then resume the paintings.

If by this method of application the effect aimed at by the doctor is too slight, then he will use the pure tincture of iodine. Too strong an effect is reached when the skin not only smarts painfully where painted, but the surrounding parts swell and become tender after the first three paintings. If the effect be more than the doctor desires, the paintings must be discontinued, and cold compresses must be applied to the swollen tender skin.

The liniment of iodine (British Pharmacopœia) is applied to produce quickly an acute superficial inflammation of the skin, with formation of blisters, as by blister-plaster. Iodine is rarely used for this purpose in Germany.

Collodion is frequently applied with a *brush* over strips of court plaster, or gauze, to fasten them to the skin. It is a viscous fluid, the ether in which *quickly evaporates* in painting, leaving a thin, shiny, closely-adhering film, which does not wash off but may be rubbed or drawn off as a whole. Collodion will not adhere unless the skin is thoroughly dried before painting; in drying, collodion strongly contracts, and therefore it is also used to produce moderate pressure, as for instance, on slightly reddened parts of the skin, and on *fungus hæmatodes* * in children. (Its use in bedsores I do not advise.) Sometimes it so contracts the skin that small blisters arise by the side of it; care must be taken against thickly painting, ring fashion, with collodion (a finger for example), as it so compresses the skin that the finger-tip becomes blue, painful, and benumbed.

Occasionally medicaments are dissolved in collodion (iodine for instance), to combine a dispersing and compressing action. *In using collodion, care must be taken not to bring it too near to a light, as it ignites easily.*

B. Applications with a brush upon mucous membranes. Paintings are ordered for the gums, the tongue, and the mucous membrane of the throat; when in the conjunctival sac of the eye, the doctor generally operates. Into the nose soft ointments are conveyed on a camel's-hair pencil. Small and large soft brushes, water-colour brushes, are used, or a brush may be made of charpie firmly fixed to a stick.

In order that the medicament (borax, nitrate of silver, or hydrochloric acid), dissolved in honey as painting syrup, or in water, shall adhere firmly to the diseased parts of the mucous membrane of the mouth, it is advisable to dry them first, then paint, and let the mouth be held open a while so that the fluid infiltrates the mucous membrane. To be effectual, the paintings must be frequently repeated.

In painting the throat in long-continued catarrhs (in which

* "A soft, malignant growth, exuberant and highly vascular, and therefore peculiarly liable to bleed."—QUAIN.

solutions of alum, nitrate of silver, permanganate of potash, and also tincture of iodine are used), the pencil holder must be lengthened.   To remove the adhering mucus let the patient first gargle with water, then, with the eyes fixed upon the point to be painted (for instance, one of the tonsils), convey the pencil direct to it, quickly touch it with the medicament, and as quickly withdraw the pencil. Retention of the brush in the back of the throat produces retching in most persons, and even vomiting.   On painting the throat in diphtheria, see page 226.

To convey soft ointment far into the nose, place it on the brush, and insert the brush horizontally into the nostril to a depth of from $1\frac{1}{2}$ to 2 inches, the head being held upright; whilst compressing the nostril, remove the brush, and let the patient draw in the ointment which has been applied. The action must be quick so that he does not sneeze, and in sneezing, expel the ointment.

When it is desired artificially to impart an oily coating to dry skin, to make it pliable and less tense, or to introduce into the skin medicaments through the medium of fatty substances, it is *embrocated with ointments, lard, or oils.* To embrocate without pain to the patient, put the ointment or oil upon a small piece of flannel or lint rolled together like a ball, and rub it in, with moderate pressure, at the place indicated by the doctor. Rubbing on inflamed parts must be done very gently, or the skin becomes too much heated and irritated. With some substances, such as mercury,* iodine, belladonna, &c. (often mixed with fats as ointments), the too frequent application, or too large a quantity of the

* If mercury comes in contact with gold rings, from amalgamation, they become white like silver. By vigorous rubbing with leather and polishing powder the quicksilver amalgam soon disappears.

embrocation, may be injurious locally, or may operate as a poison generally ; the doctor must therefore determine the quantity to be used, and the frequency of its application. After every application the nurse must at once carefully wash her hands. Parts anointed with grease must always be covered afterwards with linen, lint, or cotton-wool wadding ; the two last to be fixed by a bandage, or the clothes will be soiled with the ointment, and the stains are very difficult to remove.*

In *applying alcoholic liquids and liniments externally*, another method is employed. The object generally is, strongly to irritate the skin until it is reddened, and until the patient feels a sensation of warmth and smarting at the place embrocated. Pour some of the liquid upon a rather large piece of flannel, and with it rub the skin firmly with strong pressure until the desired result is obtained.

In fact, the rubbing is just as important in these spirituous embrocations as is the substance used ; neither, however, operates upon the skin to any considerable depth unless anodynes, such as chloroform, morphia, or veratria are combined with the spirit or liniment (mostly in oil-emulsion) rubbed in.

Friction, stroking,† *kneading, percussing,* alone, without medicaments, have been developed into a system of treatment termed " massage." The very old method of " shampooing " (never quite out of use as a popular remedy), is again revived, and, in suitable cases, proves very efficacious. But much

---

* Nitrate of silver stains may be removed from linen by using strong solution of iodide of potassium, or, still more easily, by a solution of cyanide of potassium.

† Stroking, *i.e.*, " Effleurage, the form of massage which consists in gently rubbing the surface with the palm of the hand, the direction being towards the centre of the body."—MAYNE.

injury may be caused by it.   Massage is used to disperse ex-
travasations,* fluid or coagulated exudations,† and inspissa-
tions *(infiltrations ‡)* mechanically, and thus to facilitate their
absorption, and to strengthen and harden such parts of the
body as have become relaxed and very sensitive.   For both
purposes, the utmost excitation of the minutest blood-vessels
(capillary vessels) is necessary; and this, not only that more
blood shall circulate through the parts massaged, but that
the walls of the blood-vessels shall become more perme-
able, and the tissue particles more energetic in their vitality;
in short, that metabolism § shall be promoted by the mechanical action, and thus the abnormal deposits effused into the
tissue substances will be dispersed.   To produce these effects
by massage, power of endurance is requisite; in chronic
cases, courses of massage, of months or of years' continuance,
are necessary.   Whether the course shall be continued or intermittent : several sittings daily or a given number weekly;
which method of massage, and whether it is to be applied
gently or vigorously, can be decided by the doctor only after
long experience in the use of massage in its various modifications.   Therefore, it is an error to suppose that any one
can massage successfully who has heard the treatment described, and knows the manual exercises, but who has neither
theoretically studied the subject, nor practically exercised the
system.   On the whole, few female nurses will acquire sufficient strength in the fingers and arms to carry out courses of

* Extravasation : "Effusion of a fluid, or its state when effused, and so
out of its proper vessel or receptacle."—MAYNE.

† Exudation : "Oozing of the *Liquor sanguinis* through the vascular
walls ; also, its fibrinous portion, when coagulated on the surface, or in
the substance of any tissue or organ."—MAYNE.

‡ Infiltration : "A straining of fluid substances . . . into cellular
tissue."—MAYNE.

§ Metabolism : Change of substance.—TR.

massage; even very strong men feel exhausted after massaging a great part of the day. Instruction can only be given simultaneously with the practical application of the method.

With *gymnastics* the procedure is very similar. Gymnastic exercises for restorative purposes can be practised by women, but not every woman has strength, endurance, patience, method enough for these rather fatiguing exercises.

The modes of applying *electricity* can likewise be learnt by women when the doctor specifies definitely, in each case, the intensity of the current, and the duration and frequency of the sittings, but the nurse must have some knowledge of the apparatus, so that she may quickly remedy any little defects in it.

5. LEECHES. MUSTARD PLASTERS. BLISTERS.

Formerly various methods were used for drawing blood locally; but now that by leeches is almost the only one in use, and this very seldom. Two kinds of leeches are used: the grey or German, and the green or Hungarian. In buying, care must be taken to see that they are healthy and full of life, will draw themselves together into an oval form on being touched, and that they have not been previously used. When to be applied, the leech is placed in a glass tube about half-an-inch in diameter, and 4 inches long. Test tubes as used for chemical experiments are suitable. After the place has been washed with a clean sponge, place the mouth of the glass against the spot where the leech is to bite. If it does not bite quickly, touch the point to be bitten with milk, a solution of sugar, or blood, then let it suck until it falls off of itself. If the little wound must continue to bleed, lay on clean sponges which have been dipped in lukewarm water; to stop the bleeding, dip the sponges in cold water

and hold them for a time firmly against the bleeding places.
If, in spite of this, the blood still flows freely from the wound,
a small piece of *amadou* * may be pressed upon the wound
and allowed to adhere.   For some days the little wounds must
be protected against rubbing and dirt by linen fastened by a
bandage.

Mustard plasters, or poultices (Sinapisms).   Ground mustard
(bruised or ground seeds of white or black mustard) is stirred
on a plate to a thick paste with cold or tepid water.   This is
spread thickly with a knife upon a piece of fine linen or gauze,
and covered with only one layer of the thin material.   The
plaster is laid upon the skin and left until rather intense
smarting is produced.   The skin should only be made very
red—blisters must not be raised.   How quickly the desired
effect appears depends upon the freshness and strength
of the mustard, and the more or less sensitiveness of the
skin.   Much redness may appear in five minutes, or the
poultice may fail to operate after having been applied for a
quarter of an hour.   It must, therefore, be repeatedly lifted to
see if the desired redness is produced ; if produced, remove
the plaster.   Its strength may be intensified by adding vinegar
and grated fresh horse radish, or it may be weakened by
adding ordinary flour.   " Mustard leaves " are more uncertain in their action than newly-prepared mustard poultices.

If *blisters* are to be produced upon the skin, *blistering*
or *cantharides-plaster (vesicatory)* is applied.   The chemist
prepares it by mixing pounded Spanish flies (cantharides)
with wax, turpentine, and oil.   As the vesicatory does not
adhere firmly, it is best spread direct upon the adhesive side
of a piece of prepared adhesive plaster, sufficiently large

* Amadou: "A substance (prepared from a fungus on oaks and beeches)
used for graduated compresses, to support varicose veins, abraded
surfaces, &c., and to stop bleeding."—MAYNE.

to extend three-quarters of an inch all round outside the vesicatory. On adhering, the blister rises under it, filled with water, in from eight to ten hours (with sensitive skin, somewhat more quickly). The pain caused by the drawing is insignificant, so that the blistering plaster can be applied in the evening without disturbing the patient in his sleep, unless he is hyper-sensitive.

When adhesive plaster is not used the vesicatory is spread upon linen or strong paper, and is fastened to the skin by strips of adhesive plaster, or by a bandage.

When the blister is formed and the plaster removed, the bladder is pricked with a clean sewing needle for the water to ooze out. Clean cotton-wool is then applied, which subsequently falls off with the dried-up blister. Should the skin on the blistered part slightly weep, apply zinc ointment.

6. BATHS.

Baths are divided into :

Full-baths.—The bather lies in the bath with his body half-raised, the water reaching to his neck. The bath should hold from 55 to 66 gallons of water. Baths for children require from 22 to 44 gallons.

Demi-baths.—Position of the bather as before, with water reaching to the navel only. Quantity of water, from 33 to 44 gallons.

Hip-baths.—Specially shaped baths, in which the bather sits with the thighs bent towards the upper part of the body, and the legs, from the knees, outside the bath. Quantity of water, 11 to 13 gallons.

Foot-baths.—The water reaches to the knees — special baths are necessary. Quantity of water, from 4 to 6 gallons. With ordinary foot-baths the water only reaches the ankles ; vessels for such baths may be found in every household.

Arm-baths.—The hand with the whole forearm up to above the bent elbow should be in the water; special arm-baths are necessary. Quantity of water, from 3 to 4 gallons.

Hand-baths.—The hand alone to be in the water for a time; every large wash-hand basin is suitable.

Baths are now almost exclusively composed of water, to which medicaments can be added as prescribed. The bouillon, milk, or wine baths, formerly much used and very costly, are discontinued.

Baths are either *hot*, *tepid*, or *cold*. In all German speaking countries it is customary to measure the temperature of water, as the temperature of air, "according to Réaumur." To measure the temperature of bath-water it is not enough superficially to dip the thermometer into the water—it must be done as follows: first, well mix the warm and cold water with the arm, or with a large wooden staff made for the purpose, then dip the thermometer into it, and wait until the mercury ceases to rise, when read off the temperature whilst the mercury bulb is immersed. Some *bath thermometers* are fixed in pieces of cork so as to float upright in the water. Their use is much to be recommended when baths are often taken in the house.

Hot baths should be of the temperature of $99.5°$ F. ($= 30°$ R. $= 37\frac{1}{2}°$ C.), corresponding to about the blood-heat of a healthy person, and should be used only when specially medically prescribed. If at the beginning $99.5°$ F. be too hot for the patient, begin with $95°$ F. ($= 28°$ R. $= 35°$ C.), and gradually add hot water until the desired temperature is reached. If in such a bath the patient gets a very red face, feels tension, throbbing in the head, and drowsiness, lay a cold water compress upon his head; then, if these symptoms do not soon subside, the bath must be suspended.

Tepid baths are of the temperature of from 90·5 to 95° F. (= 26 to 28° R. = $32\frac{1}{2}°$ to 35° C.). *Cold baths* may be as low as 65·75° F.(= 15° R. = $18\frac{1}{2}°$ C.).

The air-temperature of a bathroom should not be less than 65.75° F. For baths, again and again the caution must be given, *not to exchange the temperature definitions* "according to Fahrenheit," "according to Réaumur," and "according to Celsius"; often, from mistaking the one for the other, accidents, even death itself, have occurred.

The *duration of the bath* generally depends upon its temperature, unless special directions are given by the doctor. Cold and hot baths should not be continued more than five minutes ; tepid baths are generally continued a quarter to half-an-hour, and longer, especially when medicated. To prevent the water from cooling too quickly, a thick woollen blanket is laid over the bath, so that only the bather's head remains uncovered ; from time to time small quantities of hot water are added.

Weak, helpless patients *the nurse must assist, both in getting into and out of the bath, and in drying and dressing ;* children are lifted in and out. For adults confined to bed, the bath is placed by the bedside, so that the patient can go directly into it. Contact of the body with the wood or metal of the bath is very unpleasant to many patients ; in such cases, lay a linen sheet in the bath (as is always done in the bathing establishments of Vienna) ; patients suffering from bed-sores must have an indiarubber circular-centre cushion placed under them. To lift patients, unable to help themselves, into and out of the bath, the following is recommended : if the bath is to be managed by *one* person, then, in the first place, the water is prepared in the bath, and a clean linen sheet is laid in so that it can be fastened by the corners to rings, which should be at the outside ends of the bath; the tension of the sheet must be such that, like a hammock, when the patient is laid thereon, it is not pressed lower than half the depth of the bath. Of course, there must be sufficient water so that, in this position, he is covered up to the neck. He is lifted (as described, p. 72) from the bed on to the sheet, and after the bath into the bed again. To lay him carefully on to the floor of the bath, and to lift thence, is only possible to a very strong person. If several persons assist, then he may be lifted by the linen sheet into the bath, the sheet being grasped by its upper and lower ends (it is always advisable to let a third person support the pelvis and upper part of the body). In the same way the patient is lifted from the bath. In practice the first method is soon found, not only the more simple, but the more agreeable to the patient. He is not to be left alone when in the bath, because he might become faint and get his face under water.

Whilst the patient is in the bath the nurse arranges the bed, spreading over it a large blanket; on leaving the bath he lies upon this blanket and is wrapped therein. If he is to perspire (transpire) after the bath, several blankets must be laid upon him, and he must be kept quiet. After the perspiration has continued for half-an-hour, remove the extra blankets, uncover the upper part of his body, and first rub it dry with a Turkish towel; then put on a warmed vest or shirt. By degrees the remaining parts of the body are uncovered and dried, the blanket is withdrawn, and the patient is made comfortable in bed. As, after a warm bath, many patients are rather fatigued, give them a cup of bouillon or some wine; comfortable refreshing sleep generally follows. The nurse must therefore take care that the room is somewhat darkened, and the patient kept undisturbed.

Medicated Baths.—All medicated additions, referred to specifically by weight or by measure, *are for* use in *full baths for adults ;* for *children's baths,* take one-half ; for *foot and arm baths,* one-fourth.

Salt baths.—From 2 lbs. to 6 lbs. of common or sea-salt, or a corresponding quantity of mother-water,* the quantity of salt in which varies considerably.

Moor baths may be artificially produced by stirring Franzensbad ferruginous peat-earth in hot water. The earth is exported from Franzensbad (Bohemia) in sacks. For a full bath, about a hundredweight of moor-earth is necessary. The artificial moor baths are generally used as hand or foot baths, for which from 13 to 22 lbs. suffice.

Moor extracts (moor salts and moor lye) are now articles of commerce, and are substitutes for moor-earth. If *poultices of moor-earth* are prescribed, then flat bags, of a size to cover the affected part, are half-filled with the earth, the ends sewn together, and the bag dipped into hot water every two hours ; the water is drained off, and the wet poultice laid on like a cataplasm. As moor-earth stains are very difficult to remove, care must be taken that the linen is not soiled by the earth.

Ferruginous baths may be produced artificially by the *globuli martiales,* which can be prepared by the chemist ; each ball weighs about 1 oz., and from one to three such balls are dissolved in the bath water.

Alum baths are only used as hip baths ; $\frac{3}{4}$ oz. crude alum suffice for a bath.

Sulphur baths are usually made by adding to the water from $1\frac{3}{4}$ ozs. to $3\frac{1}{2}$ ozs. of sulphurated potash *(Kalium sulphuratum pro balneo).*

---

* Mother-water : "The residual fluid after evaporating sea-water, or any other solution containing salts, and taking out the crystals."—MAYNE.

Aromatic baths are thus prepared : about 2 galls. of water are poured boiling upon 2 lbs. of *Species aromaticæ* (obtained at the chemist's) ; the infusion stands covered for a quarter of an hour, is then strained through a cloth and added to the bath. Many chemists prepare aromatic spirits for baths, of which from 2 to 4 fluid ounces are added to a bath. For *camomile* and *calamus baths*, prescribed for children, from 8 ozs. to 16 ozs. of dried camomile flowers, or of calamus roots cut in pieces, are thrown directly into the bath-water, or a strong infusion of one or of the other, previously made with boiling water and strained, is mixed with the bath water.

Pine-needle baths are artificially made by adding extract of pine-needles (or extract of fir-needles). The method of preparing these extracts varies ; more must be added of the weak than of the concentrated extracts—a general rule cannot be given. Directions for use are always affixed to the bottles containing the extracts.

Malt baths are much used for children. For a child's bath, from 2 to 4 lbs. of bruised, dry malt are boiled for half-an-hour in from 1 to 2 gallons of water, and the liquid is strained through a cloth and poured into the bath. If beer-wort from a brewery can be obtained instead, then from 1 to 2 quarts added to the bath will be sufficient.

Vapour baths require special arrangements, which cannot very quickly be fitted up in a private house. The patient is first rubbed and massaged in a room filled with hot steam, then he passes under a cold shower bath or into a cold full bath of short duration, and finally, is enwrapped in a blanket until heavy perspiration supervenes, after which he is rubbed dry. This is rather a severe strain upon the body, and should not be undertaken unless medically prescribed.

7. AFFUSIONS (DOUCHES). DAMP FRICTIONS. PACKINGS. LOCAL WET DRESSINGS.

All these processes are carried out with cool or tepid water and form essential elements of the cold water treatment; to these, the very important limitation of diet, much open-air exercise, and restriction of drinks to water must be added.

Affusions *(douches)* may be used in different ways and for very different objects. In all cases in which patients are not accustomed to the sudden influence of cold water, it is advisable first to wash the chest, armpits, and head with some of it, so as to get them used to it. Affusing the whole body with large quantities of water from a bucket can only be done whilst the patient stands or sits in a bath. For an unconscious patient lying in bed, affusions of the head are sometimes ordered; for these, let him lie, undressed, upon his stomach, with his head held over the side of the bed, which is protected by india-rubber sheeting; a large, shallow bath is placed to receive the water, and the water is slowly poured from a jug upon the head from a height of about a foot. Length of time for the affusion must be ordered by the doctor.

Similar to the affusion, but much stronger in action, is the *jet-douche.* In this the water, with more or less force, issues from an opened water-pipe, and the jet is directed on to the part of the body as ordered; this arrangement can only be used in specially constructed bath-rooms.

Shower baths are most used: they operate differently, according to the height from which the water falls, and its temperature; such douches can be fixed over a bath, or in a bath cabinet. If the shower bath is to affect only an arm or

a knee, then a shallow bath is placed underneath it, the rest of
the body being well protected by waterproof material. From a
water-can, to which a rose is affixed, held about a foot above
the diseased part, the water is allowed to fall upon the limb
for from three to five minutes.

After the douche, rub thoroughly dry with a Turkish towel.

Damp friction is frequently used to harden the skin,
especially with persons who suffer much from catarrhs and
rheumatism, and is applied in the morning directly the
patient gets out of bed. A large, coarse linen sheet, dipped
in water, is wrung out, and, by the two corners of one end,
is held by the outstretched arms before the patient, who
had previously taken off his shirt. The sheet is then wound
round him up to the throat and over the arms, so that it
cannot slip down. The body is now vigorously rubbed up and
down over the clinging wet sheet for several minutes; after
which a large, dry, coarse sheet is wrapped round him, and the
rubbing repeated until he is dry. Anæmic persons, who, in
the morning even in bed are not very warm, must be warmed
artificially before the damp friction is applied, either by lying
for a time under thick blankets (but without perspiring), or
by being previously dry-rubbed until the skin of the whole
body has been made warm and red. Then the chest,
shoulders, and stomach are wetted, and the patient wrapped
in the sheet, previously dipped in tepid water and well wrung
out. In this way the effect of the friction is somewhat
weakened, but by such means it is possible to apply it to
persons who cannot endure severe cold.

For *packing* or *swathing* in wet linen sheets, spread a large
blanket over the bed, lay on it a linen sheet (previously
dipped in cold or tepid water as prescribed by the doctor,
and well wrung out); on this lay the patient and wrap it round

his body, each arm and leg separately, until the whole body is enveloped ; then closely wrap the blanket around the body, leaving the head free. Usually the object of packing is to produce perspiration ; this can be greatly assisted by raising the temperature of the room, and by laying on several blankets. If cooling the body is the object, the patient must not remain too long in the packing. After the packing, rubbing dry follows.

Wet dressings of different parts of the body are frequently ordered (hydropathic compresses, Priessnitz ditto, stimulating ditto). For these, a sufficiently large towel is folded lengthwise, dipped in water of the temperature of the room, wrung out, and wrapped evenly around the affected part (throat, chest, abdomen, knee, elbow, or hand) ; then a piece of waterproof material (or even a thick piece of flannel) is so laid as somewhat to overlap the wet compress, and the dressing is fastened by handkerchiefs or bandages. Such a compress is renewed every two hours.

8. CONTINUOUS APPLICATION OF COLD AND OF HEAT.

1. *Cold.*—To continuously keep a diseased part of the body at the lowest possible temperature, it will not suffice to lay upon it compresses dipped in water from time to time ; this slightly cooling influence may be adequate in many cases, but, if strong effect is desired, ice must be used. For this purpose an india-rubber bag, filled with crushed ice, is laid upon the part affected. Much depends upon the way in which this is done. Above all, the ice-bag must not be laid next the skin ; first cover the skin with a four to six-fold layer of dry linen, or insupportable pain may be produced in it by the severe

cold, or even frostbite may occur. The form of the ice-bag must be carefully selected—it must fit close to the part to be treated, and great care must be taken that it is securely closed. Even by tightly tying the bag-opening with string or tape a waterproof closure is not always obtained, and the india-rubber may be easily cut. A bag of natural or vulcanized india-rubber, having a broad open mouth and without special closing appliances, is best closed with a shallow wooden stopper (about $1\frac{1}{4}$ inches in depth); this fits into the opening, and has somewhat curved sides, so that the rubber can be firmly secured around the stopper by a broad tape. Proper clamps for closing ice-bags can be obtained. To prevent the ice-bag from slipping off the affected part it must be secured to it by a bandage. If the part of the body treated is already bandaged (for instance, a broken limb), then, to prevent the weight of the ice-bag from being burdensome, suspend it from a curved rod over the limb, so that the whole weight does not rest upon the limb. Two ice-bags should always be provided: when the ice in the bag in use is melted, the second, previously filled, is substituted for it.

In filling ice-bags the following rule should be observed:— Wrap the ice in a coarse clean cloth, lay it on a firm basis, and, with a hammer, break it into pieces about the size of walnuts. This should be done some little time before filling the bag, so that the sharp edges may be melted off. (Pieces of ice may be broken smaller by pricking with an awl, bodkin, or a very strong pin.) The ice-bag is to be half-filled only; in closing, press the bag firmly around the ice to expel the air.

Difficulties arise when ice is to be applied to the back of a recumbent patient, or to the underside of a limb, viz. : if a part of the body is laid *upon* an ice-bag, it will be found, that the air in the bag, and the ice melted to water—which are both in the upper part of the bag—become so warmed by the part of the body lying upon it, that the ice no longer exercises any cooling effect. In this case care must be taken that the air and water in the bag may escape. To effect this an elastic catheter is fastened into the bag, to which an india-rubber tube is fixed and conducted from the bed into a vessel underneath.

In emergency, india-rubber bags can be replaced by pigs' bladders, or by bags of vegetable parchment; the former soon become offensive, and the latter seldom continue perfectly watertight.

To keep ice upon the head of patients suffering from injuries to the brain, or from inflammation of the brain, caps of india-rubber are made on each of which there is a bag for ice. When necessary a bag can be stitched upon a woman's nightcap, into which the ice-bag is put, and the bag itself is then drawn together like a tobacco-pouch. When patients are very restless, there is great difficulty in fixing the ice-bag upon the head, as it falls from side to side—then it must be as repeatedly re-placed.

If the parts to which cold is to be applied are very sensitive, occasionally even the small crushed ice cannot be tolerated, because of its weight. Then towels, folded together and laid upon large blocks of ice, are cooled, are applied to the inflamed part, and are changed every three to five minutes. Instead of ice-bags, refrigerators * are now much used. These are of various forms (mostly in the shape of plates or caps) adapted to the part of the body to be cooled. They consist of slender tubes through which ice-water continuously flows : thus a pipe conveys the cold water from a vessel standing

* Leiter's Coils.   Patentees, Krohne & Sesemann, Duke Street, London, W.—Tr.

high by the side of the patient, to the refrigerator, and another pipe conducts the water, warmed by the body, into a vessel on the floor. These appliances are made of india-rubber or of pewter; the former is more soft and pliant, but the latter permits more intense effect of the cold. The apparatus is firmly fastened to the part of the body to be cooled, and remains unmoved; care is necessary to see to the supply and off-flowing of the water, and thus the patient will be very little disturbed.

Cold may be applied continuously in two other ways—one, in the form of *continuous cold water baths*, in baths made specially for the foot or the arm, placed in the bed; the other, in the form of *continuous cold irrigation* (douching). For such applications, special appliances are necessary; as a rule, they are only kept and used in hospitals. Their management must, in each case, be explained to the nurse.

2. *Heat.*—Heat also can be continuously applied in the form of *continuous warm baths*, but *hot compresses* are more frequently used. The simplest way is, to dip a folded towel into hot water, wring it out, and lay it on the affected part; cover it with a piece of waterproof material that over-laps the compress, fasten with a cloth or bandage, and renew every half-hour or hour. If the patient is to be spared the process of renewal, then, to keep it warm, the compress must be covered on the outside with a dry hot compress (bag of oats or bran, see page 130), which can be more easily changed. In most cases these *hot water poultices* suffice, but it is generally true that, by the older method of preparing poultices, the warmth acts more equally and is more lasting. In preparing *cataplasms* (poultices) according to the old style, a pap of linseed meal, groats, or bread is used. Boil the meal or groats in water to a thick pap, and keep it standing on the hot

stove or hot hearth ; with a spoon spread the pap on a large piece of linen, quite enclose it therewith, and then lay it upon the place to be poulticed. To keep the pap hot, and yet prevent it from burning, take a tin saucepan, not too deep, lay in it two little pieces of wood, upon which place the basin of pap, and pour hot water into the saucepan. If this improvised *bain marie* is placed upon the fire, the pap can be made very hot without burning it. If the kitchen be distant from the sick-room, the pap poultice may be prepared for use, be laid upon a hot-bottle filled with hot water, and then taken to the room.

In many places the people prefer to *soak bread* (white or black) in hot milk, and to lay it, wrapped in linen, upon the affected part. The addition of onions or of honey is said to · draw pus from inflamed parts very quickly—a harmless popu- lar tradition. For persons who believe that they will not recover unless something is fetched from the chemist, the doctor may perhaps prescribe *Species emollientes*, which are stirred to a pap with hot water in the same way as meal or groats.

When applied, it is of the greatest importance that these poultices shall not be too hot, because they often produce slight scalds. Decision by the thermometer is not requisite ; before applying, the nurse must hold the poultice to her own cheek, to test if the heat be bearable.

In preparing and applying cataplasms, great carefulness and cleanliness are essential. Formerly, burnt and soured poul- tice-pap diffused a very peculiar odour through the hospitals. A cataplasm, well covered by waterproof material with flannel laid over it, keeps sufficiently hot for two hours ; when removed, a fresh one must be ready to lay on.

By long-continued application of damp heat, particularly to

the hand or the foot, the great swelling of the thick epidermis often causes very unpleasant dragging pains.   To prevent this, first anoint the affected parts with almond or pure olive oil.   Long application of moisture may cause little nodules, and even small vesicles, to arise on tender skin ; anointing in the same way will prevent these also.

In many cases (for instance, for soothing colic pains) the *application of dry heat* is quite sufficient ; for this purpose, bags of bran, or of crushed oats, which can be quickly heated on a not too hot stove, or in an oven, are best.   For the same purpose, so-called "hot-stones" are made.   These are generally marble slabs about three-quarters of an inch thick, slightly concave, large enough to cover the stomach ; one of these is heated like the bran bags, wrapped in flannel and applied, or india-rubber ice-bags filled with hot water may be used.

CHAPTER IV.

PREPARATIONS FOR OPERATIONS AND BANDAGINGS.

Antiseptic (disinfecting) Remedies.
Preparations for Operations. Preparation of the Nurse.
Selection of Room. Lighting. Preparation of Patient.
Temperature of Room. Operation-Table. Anæsthesia.
Directions for Making Sponges and Silk Clean Surgically
(to disinfect). Names of most-used Surgical Instruments.
Preparation for Antiseptic Dressing of Wounds. Tam-
pons. Plaster. Dressing of Wounds at the Sick-bed.
Preparation of Solutions of Carbolic Acid. Salve-Dress-
ing. Preparations for Bandaging Fractures, Dislocations,
Curvatures. Plaster of Paris Bandage; Starch ditto;
Silicate, or Water-glass ditto. Gutta-Percha Splints.
Practice in applying Roller and Handkerchief-Bandages,
with Illustrations. Transport of Patients.

ONLY upon very able female nurses, whose absolute reliability and carefulness have been often tested, can the doctor depend—I might almost say, as a special distinction—for direct assistance at operations and in dressing of wounds. But *all* trained female nurses should understand how to prepare for operations, should know the most-used surgical instruments, and how to get ready everything likely to be needed in the different bandagings or dressings.

In practical hospital service this knowledge will be speedily gained by an able woman, of quick perception, and keen observation. From oral lectures with demonstrations much also may be learnt in a short time that even a very detailed explanation could scarcely render completely intelligible.

Therefore I shall now state only that which the female nurse, by frequent reading, must impress upon her memory.

ANTISEPTIC (DISINFECTING) REMEDIES.

Antiseptic remedies, that is, those remedies which are adapted to destroy disease-germs, are very numerous. As each, with its advantages has also its disadvantages, thus rendering it unsuitable for certain purposes, so one alone does not suffice for all purposes. In order not to perplex the reader by referring to too many antiseptics, only a few will here be described.

1. *Carbolic acid* is used in aqueous solutions to disinfect the hands, instruments, the wound itself, and the skin surrounding it, the sponges and tampons, and the ligatures for sewing of wounds and for tying bleeding vessels; it is used also for carbolising the materials for dressings.

The aqueous solutions are thus prepared: pure carbolic acid as sold in solid form in dark glass bottles, is liquefied by heat (by placing the whole bottle in a vessel of warm water); the quantity to be used is determined by weight, or by measuring-vessels specially gauged for the purpose. For solutions of 1 per cent., 175 grains; for 2 per cent., 350 grs. ; 3 per cent., 525 grs. ; 4 per cent., 700 grs. ; 5 per cent., 875 grs. of the carbolic acid are added to a quart (Imperial) of water, and are dissolved by being well shaken. Bandage-gauze (very loosely woven mull) is carbolised as follows : mix by stirring, 1 lb. of finely-powdered resin in 5 fluid ozs. of alcohol ; 4 ozs. of carbolic acid ; 3 fluid ozs. of castor oil (or instead of the oil, 4 ozs. of melted stearine), and keep this mixture in stoppered bottles ready for use. When used, stir this mixture into half-a-gallon of alcohol ; open out and disarrange the muslin, place about 2 lbs. weight of it in a shallow bowl, and pour the solution over it; when saturated, wring out and hang it up to dry for from 5 to 15 minutes, according as the room is cool or warm. Thus prepared, it is preserved in closed tin cases. Carbolic acid is poisonous; in stronger solutions (5 per cent.) it is corrosive. As it is very volatile, the solutions and the carbolised gauze must be kept in air-tight vessels.

2. *Perchloride of mercury is used in aqueous solutions* 1 : 1000, ($17\frac{1}{2}$ grs. to 1 quart of water), to 1 : 5000 (14 grs. to 1 gallon of *water*), in the same way as the aqueous solutions of carbolic, but not for disinfecting metallic instruments, as it corrodes them ;—therefore, these solutions must not be poured into metallic vessels, cups, irrigators ;—in *alcoholic solutions* of 1 : 1000 (if by weight, $14\frac{1}{2}$ grs. to 1 quart of *alcohol)* for sponges, tampons, and ligature-threads, and for impregnating dressing-gauze. The gauze is thus prepared : in 5 fluid ozs. of glycerine and 1 gallon of water, mix 70 grains of perchloride of mercury, and 1 lb. of common salt, and in this solution soak the gauze ; when saturated, well wring out and dry.

The aqueous solutions are mostly prepared as follows : one pellet of perchloride of mercury (containing $15\frac{1}{2}$ grains of perchloride of mercury combined with common salt or tartaric acid) is dissolved in the necessary quantity of water (1 quart to 1 gallon). Perchloride of mercury is very poisonous.

3. *Iodoform* is used in crystals, or powdered (as sold by chemists), and for making iodoform-gauze. The gauze is thus prepared : mix 3 ozs. of iodoform, 30 fluid ozs. of alcohol, and 10 fluid drachms of glycerine ; when mixed, pour it over 10 yards of gauze, and then rub as in washing of linen, until the whole is equally saturated. Iodoform (profusely used) acts as a poison, and has a penetrating smell.

4. *Salicylic acid* is used in solutions, of 40 to 50 grains of crystallized salicylic acid in 2 pints of water; and as salicylic ointment (1 : 10, or 45 grs. to 1 oz. of vaseline) for rubbing into the skin against eruptions, frequently caused by carbolic acid, and by perchloride of mercury.

PREPARATIONS FOR OPERATIONS.

First of all, the nurse must prepare herself and her clothing; she must dress her hair and firmly fasten it; then cleanse her hands and forearms to the very cleanest with soap and nail-brush. For every operation she must put on a fresh, clean, linen overall, fastened at the top close round the neck, and buttoned in front from the throat down to the bottom of the dress. The sleeves must come to within a handbreadth of the wrist, and be hook-and-eyed together, or be closed by an india-rubber band. The fresh, clean overall, fresh for every operation (the operator and his assistants being similarly supplied), is to protect the wound against injurious matters which might have settled on the clothes of the nurses, or of the doctors, in their intercourse with other patients, and also to prevent the clothes from being soiled.

The room in which more important operations are to be performed must be well lighted. It should have wide windows on one side only, extending upwards as high as possible. If urgent operations must be performed at night, then it is most desirable to have the whole room brightly lighted by several gas-burners. One gaselier only, just over the operation-table, does not suffice, as dark shadows are formed beneath it, which are often very disturbing. If gas cannot be had, then use several small lights suitable for the purpose. But one must be certain that the persons holding these lights will, under no circumstances, faint away during the operation. A very good light may be obtained by taking from 4 to 6 pieces of wax-taper about 16 inches long, twisting them together firmly, and lighting one end of the twisted wax-torch thus made ; several such torches give a very bright light, and, though they burn rather quickly, they have this great advantage over ordinary wax and stearine candles, that they not only burn far more brightly, but they can be bent at will to give the light just where the operator needs it. Persons holding these lights must take care not to burn the patient, the operator, or his assistants, nor to let the hot wax drop upon them. To prevent this, a piece of thin pasteboard or stout paper, with a hole in the middle through which the torch passes, is sufficient. If, in addition, a large concave mirror be held behind these wax torches (for which a shining china or metal plate will serve), the field of operation can be almost as brightly lighted as by daylight.

Antiseptic liquid, abundant water for washing, and a large vessel to receive the waste, must be provided. For the operator, and for each assistant, two towels and one wash-hand basin each, with an abundant supply of water ; a can of hot water and two large vessels in which the sponges may be cleansed ; two large basins containing antiseptic liquid (for instance, solution of carbolic acid, or salicylic acid, or perchloride of mercury, as the doctor may order), one, to dip the hands in immediately before and during the operation, the other, in which to lay the instruments. A shallow bowl or basin must be ready, should the patient vomit during the chloroform narcosis.

Formerly many preliminary arrangements were made *to prepare the patient for an operation*, but few of these are now deemed necessary. If not forbidden by the doctor, it is always advisable to give the patient a warm, cleansing bath the day before the operation, as it probably will be impossible for some time afterwards; and an aperient should also be given. For four hours before the operation the patient should not eat, or the narcosis will most probably be interrupted by vomiting. The nurse must never excite fear in the person to be operated upon, but rather encourage him and speak of similar cases which ran their course favourably. It gives much comfort to the patients to see and speak to some one who has recovered after an operation similar to that they themselves are anticipating.

If a lengthy operation is in prospect, with considerable loss of blood, the room must be heated to 77° F. (= 20° R.; 25° C.); the legs of the person about to undergo the operation must be wrapped in blankets. The bed, prepared to receive the patient after the operation, must be warmed, and sheets wrapped round hot-water bottles (for this purpose, hospitals have small gas-stoves, which can be quickly heated) must be placed ready to warm him artificially if necessary. Hot, strong coffee, strong wine, or brandy must be at hand. The bed must be arranged according to each individual case, and a large waterproof underlay must be laid on the part of the bed corresponding to the part of the body operated on. After the operation a basin must be kept ready for some hours, as the patient may vomit in consequence of the chloroform-narcosis.

Hospital *operation tables* are rather narrow, and so arranged that, with few pillows, the patient can be placed in various positions. The table is covered with a thin mattress, cased in india-rubber or leatherette cloth, which can be easily washed. A bolster, covered in the same way, is often used to place the head in a particular position. To move such an operation table to private houses would be very inconvenient; therefore the surgeon endeavours to manage with the furniture in the house.

It is most injudicious to perform a severe operation on a patient in the bed in which he is afterwards to lie, as the bedding is seldom preserved from blood-stains and thorough saturation. Where possible, the person to be operated on should always have the full benefit of the narcosis. For this it is necessary that he should be narcotised to unconsciousness either in bed or on a sofa, thence lifted on to the operation table, kept unconscious until the bandaging is finished, and then allowed to wake in a clean newly-made bed. In hospitals, this cannot always be carried out, but it may be in private practice. The getting on to the operation table is the most painful moment for the patients : wherever possible, this should be spared them. It can scarcely require to be specially mentioned that *no preparations for the operation should be made in the presence of the patient:* he should see nothing of instruments, bandages, or other arrangements. Very timid women should be left in their morning dress, but the bands of their underclothing should be loosened for breathing easily, and they should only be undressed as far as necessary when narcotised.

A severe operation cannot be performed upon a sofa having a back : a lounge sofa without arms is better *(chaise longue),* but this is very inconvenient for the surgeon and for his assistants, because it is too low; even the strongest man in a continued stooping position gets unbearable pains in the loins, and the blood flows to his head. Such inconveniences make all who are present nervous and unquiet. If insufficient light, or other unlooked-for difficulties in the operation should be added, the person operated on will suffer for all the discomforts of the doctor by the longer duration of the operation, and by greater loss of blood. Rapidity and sureness in operating, as well as the smallest possible loss of blood, depend greatly on this: that all preparations have been so judiciously made that they prevent the failure of any incision, any ligature of a vessel, any intended proceeding, and they ensure the proposed result of every step in the operation.

In private practice, when important operations are to be performed, a suitable table is sought—a round one is quite unfitted. The best would be a table about 30 inches broad, and 6 feet 6 inches long; to this should be tied a narrow mattress with a wedge-shaped bolster. Long, modern tables are usually too broad; two small narrow tables, bound tightly together, will better supply the requisite level for the operation. If a very high position of the upper part of the body, or an oblique position, is necessary for the operation, then, in a private house, one may easily be perplexed how best to effect this. If the question be, how to secure a quiet moment, then a suitable support of pillows held by several persons will suffice, but for a longer operation with narcosis, a very firm base is requisite. This is mostly managed by placing an ordinary arm chair upon the mattress, already fastened to the table, so that the chair rests upon the front edge of its seat and the upper edge of its back, whilst the outer back of the chair is turned to the patient's back as a sloping surface; on to this a hard pillow is fastened, the chair being fixed on the table in the position described. Much time is often lost in getting the materials for such a table, and firmly fixing them so that there shall be no collapse during the operation. Everybody runs in search of one thing or another; surgeon, assistants, nurses, and others

desirous to help, get into the greatest excitement and confusion. Should the unfortunate mistress of the house be the
patient, and in her trouble must say where the linen, &c.,
lies that the doctors require, then it may be seen why the
surgeon (who may often have experienced the painfulness
of such scenes) is not prepossessed in favour of such "simple
arrangements" and "improvisations," which, in spite of the
trouble bestowed upon them, are not so serviceable as the
simplest operation-table made for the purpose.

Notwithstanding the care taken, and the covering with
linen sheets or india-rubber sheeting, the carpet or the pillows
are easily soiled with blood, or have carbolic acid, &c., spilt
upon them ; even if, *before* the operation, nothing was deemed
too precious to be sacrificed for the patient, yet *afterwards*
his friends discover, that the operation was not quite so
serious, and the doctors could very well have taken more
care so as not to have soiled the furniture and carpets.

As, in the present method of operating, very much antiseptic
liquid is used, all the parts surrounding the field of operation
must be covered with india-rubber sheets. The place itself
must first be most carefully cleansed with soap, water, and
brush (should it be covered with hair it must be shaved),
and then be washed with the disinfecting fluid.

The *sponges* to be used during the operation require most
particular care. The finest, softest, whitest sponges still contain substances which can produce very virulent inflammation
of the wounds, and may cause death. To annihilate these
substances (finest dry spores of putrefactive fungus), which
are very difficult to destroy, and to make the *sponges clean
surgically* (inferior qualities thereby also becoming white and
soft) the following is the process :—the sponges, freed from
little stones, are put into a bag and beaten until no more

sand falls out, then strained in cold water until it is no longer cloudy; this may take several hours. If to be bleached, lay them in a large earthenware vessel, and pour over them until the vessel is full, a half per cent. solution of permanganate of potash, and let them remain three hours; the liquid, now clear, is then poured away, the sponges strained, and water again poured over them, in which they remain twenty-four hours. Now they are again strained once, the water poured away, and a two per cent. solution of hyposulphite of soda poured over them. Then, from twenty to thirty drops of concentrated hydrochloric acid are added to a quart of the last-named liquid, and the black sponges are left therein until they become white—this takes about twenty minutes. They must now be quickly taken out and thoroughly rinsed. As they are quite soft and white, put them into a five per cent. solution of carbolic, or into an aqueous solution of perchloride of mercury (1 : 1000), in which they remain until wanted. Sponges already soft and white, and bleached sponges which have been used during an operation, are soaked for two days,—after repeated strainings, they are best put into a jar or large glass with water, in a moderately warm place,—and before being again used, they must be laid in the antiseptic fluid for fourteen days. The solution in which they lie must be wholly renewed twice monthly, because the perchloride of mercury, and the carbolic acid to a yet higher degree, volatilize.

Sponges which have been used for clean wounds must be carefully washed with soft soap (also called potash soap, green soap, black soap), rinsed, and again laid in the antiseptic solution. *Only such sponges should be used which, having been from two to three days in water, were laid for fourteen days in a five per cent. solution of carbolic, or in a per-*

chloride of mercury solution (1 : 1000); when used, they are taken directly out of the antiseptic solution, strained, and laid ready for the operator in a wash-hand basin containing one per cent. solution of carbolic, or a weak perchloride of mercury ditto (1 : 5000). *Sponges* used for ichorous wounds and sores, in diphtheria, and in hospital gangrene, *must be burnt immediately* after the operation. On account of the many processes necessary for disinfecting sponges, tampons are now frequently used instead, and they are thus made :— loose balls of ligneous fibre *(wood-charpie)* are made of the size required, tied up in a double layer of gauze, and kept in antiseptic solution. For rapid disinfection of tampons, gauze, and substances not destroyed by high temperatures, disinfecting or sterilizing ovens are now frequently used in hospitals. In these steam at 212° F.(= 100° C.) passes for an hour through the materials laid in them.

It is difficult, but very important, to make the *silk* used in ligaturing bleeding vessels, and for suturing wounds, quite *free from putrefactive agents* (aseptic). This is done in the sterilizing oven, or by boiling the silk, on reels, in a five per cent. carbolic solution, to which, while boiling, some more pure carbolic acid is added. A small petroleum stove, burning for about an hour and a half, may be used, and this gives sufficient time to disinfect the silk. Gas-stoves also may be used. The silk, when boiled, is placed in glass bottles having wide mouths and glass stoppers, filled with five per cent. solution of carbolic, and from these it is taken for use.

It is a proof of the greatest confidence when the surgeon entrusts the nurse with the disinfecting and keeping clean of sponges and silk. Not only rapid healing without suppuration, but the life of the person operated on frequently depends upon it. The greatest surgical skill, and the greatest care in the after-treatment are exercised in vain if, during the operation, injurious substances be conveyed into the wound by sponges, instruments, silk, or dirty fingers.

Skilful nurses may be appointed to place the instruments ready for the operation, but the requisite knowledge must be gained by practice. I will here name some of the most generally used surgical instruments.

Operation knives differ in size and form. For straight knives, the Latin term " *scalpel* " is used ; the French term, " bistouri," is applied to a knife with the blade shutting like a pocket-knife ; knives with quite straight backs and edges ending in a button or knob at the point are termed "knobbed," or, after their inventor (Percival Pott), Pott's knives.

Scissors differ in size, and are pointed or blunt as required ; they are also used with the shears curved on the flat,— " curved scissors."

Various instruments are used for grasping the parts to be cut, such as *pincettes, hooks*, and *forceps ;* these vary in size and strength. *Pincettes* at their points are either grooved (anatomical), or have one or several fine hooks *(small hooked or toothed forceps)*. *Hooks* are *single, double*, or *multiple (rake-hooks)*, and are either *blunt* or *sharp*. *Forceps* are either blunt, grooved at the points *(dressing forceps)*, or terminate in pointed hooks, and are termed *Muzeux' forceps*, after their inventor.

For temporarily holding bleeding vessels until tied (ligatured), *clamp-forceps* (clamps), or *artery forceps with a sliding catch*, are applied, which are kept closed by a slide fastened to them. To give vent to fluids from cavities of the body, a *trocar* is used ; this is a pointed, three-sided steel rod (with handle) surrounded by a close-fitting tube ; in puncturing, the tube also enters, and, after the withdrawal of the "stiletto," remains until the fluid has run off.

To destroy morbid substances by actual cautery *(thermo-cautery)*, the apparatus invented by *Paquelin* is now almost solely employed. Its essential part is a piece of platina,— this is heated in a flame, and kept red-hot by petroleum-ether and the blowing of a bellows. The nurse must know how to prepare this apparatus for use.

For cutting through bones, *saws* and *chisels* are used ; (these are divided into amputation or curved-saws, tenon-saws, chain-saws, and circular-saws or trepans).

For the examination of fistulas or sinuses, to find how deep they are, and whither they lead, flexible and knobbed rods of metal are used, termed *probes*. If a passage or a fistula is to be laid open, a grooved probe (grooved director) is previously introduced.

Longer flexible rods and tubes, of various thicknesses, are used to enlarge passages *(bougies)*, and for injecting fluids, or to let them run out. The long, slender elastic tubes with a small funnel at the top, which are introduced through the nose or the mouth into the stomach to convey fluid nutri-ment into it, are called *œsophageal or stomach-tubes*. Tubes of metal, or of elastic substances (linen saturated with a species of varnish) which drain off urine from the bladder, are called *catheters*.

By *speculum* is understood, not only the apparatus by which the reflected image of deep-seated parts is seen, as the *laryngoscope*, but also the apparatus by which deep-seated parts are made accessible to light. The latter kind of mirrors or *specula* consists of variously shaped tubes, semi-tubes, or iron plates with handles, which, by means of screws, expand *(mouth speculum).*   Such a mouth speculum should always be ready when chloroform is administered, so as, during narcosis, to be able rapidly to open the mouth, which is sometimes contracted convulsively, and to draw forward the fallen-back tongue.

For reasons given (p. 102) the use of syringes is replaced, as far as possible, by *Irrigators ;* these should never be wanting at any operation, and should be filled with water, or with antiseptic fluid, as required.

Also, at every operation, a glassful of *drainage tubes* or drains should be ready.   These are india-rubber tubes of various thicknesses, pierced with holes in the sides as directed by the surgeon.   Like sponges, these drainage tubes should always be laid two days in water, and at least fourteen days in a five per cent. solution of carbolic acid, before being used at operations.

All instruments used in an operation must *be most scrupulously cleansed* afterwards, the pincettes and saws especially must be scrubbed with a brush, and the tubes thoroughly syringed through : all must be well dried before being replaced in their cases.

ANTISEPTIC DRESSING

(Antiseptic, *i.e.*, Anti-Putrefactive).

Generally there is agreement on the principles of wound-dressing, although the materials employed may not always be the same.

Prepared dressing gauze is first applied to the wound. This is made in many factories, and can be bought almost anywhere, yet it is most important for the nurse to know how to prepare it as directed on pp. 132-3.   In many cases, a piece of *protective, i.e.,* a fine, disinfected waterproof material (oiled silk or guttapercha paper) is previously laid direct upon the wound.

The dressing gauze is laid upon the wound in disarranged curled strips *(crumpled gauze),* or in multiple smooth layers ; or in eight-fold layers which are cut of the size determined by the doctor in each case, and these, far overlapping the wound on all sides, are applied. To prevent the immediate saturation of the bandage by the blood still oozing from the wound, a piece of waterproof material, or a *wood pad, i.e.,* a cushion of appropriate form made of bandage muslin sewn together and filled with ligneous fibre, is next applied.

This dressing is so fixed by a gauze bandage that it fits closely ; all cavities and hollows are filled with dressing cotton-wool (Bruns' wadding, German charpie), or crumpled gauze, and projecting parts of bones are covered with cotton-wool.

If the dressing is to remain some time it is covered with an organdy * *bandage,* dipped in water just before use ; this dries in a few hours, and covers the dressing like a thin coat of mail. *Elastic bandages* are also used to keep the edges of the bandage lying close to the body.

To fill wound cavities, *Tampons* are sometimes used ; they are pledgets or balls of dressing wadding or dressing gauze, around which one end of a disinfected strong thread (of silk or of thread), is tightly bound, which thread hangs out of the wound cavity so that, by it, the tampons can be easily drawn out.

From its high price charpie is almost wholly discarded, but further, it has no advantage over dressing cotton-wool or gauze ; in any case, it would have to be most carefully disinfected before use.

* "Organdy (Fr. *Organdi* = book-muslin), a kind of muslin or cotton fabric, characterised by great transparency and lightness ; book-muslin as used by bookbinders."—WEBSTER.

It is very important for nurses to know how to spread adhesive plaster properly upon linen or calico, as it only adheres firmly and securely when newly-spread. The best way is, first to dip the plaster-stick in hot water, then knead and spread it equally with a strong spatula (knives are too weak) upon the tissue, held stretched upon the table by the hands. This plaster adheres much more firmly than that made with a machine by the apothecary.

In serious surgical practice *court plaster* is seldom used, yet for household use it is of great value. It is sold by all chemists—if kept dry, will continue good for a long time.

WOUND-DRESSING ON THE SICK-BED.

The changing of the dressing, often dreaded by the patient without reason, will take much less time when the nurse has every requisite prepared beforehand. Unless a particular reason exists, the dressing is only changed when blood or wound secretion has soaked through to the surface of the bandage. It is generally removed as follows: cut through the hard crust of the bandage with specially shaped scissors, or a strong sharp knife, and remove it in two or three pieces; not till the last act is the patient required to sit up, or to have his leg or his arm lifted. The removed dressing, which, in favourable cases, even after days or weeks' application, may be wholly inodorous, is provisionally laid aside. If there has been no suppuration, the fresh antiseptic dressing as described is again applied, and the nurse must prepare everything for it. Nevertheless, as it cannot be certainly foreseen whether suppuration has not taken place under the dressing, so for this also, everything must be prepared.

(*a*) One anatomical pincette, one dressing-forceps, one

scissors, one probe, must lie in a shallow basin filled with a solution of carbolic acid (2½ to 3 per cent).

(*b*) In a second hand-basin, or in a shallow bowl, with anti-septic liquid, lay cotton-wool pledgets made into balls for cleansing the wounds. Sponges should not be used at all in sick-rooms.

(*c*) An empty bowl, in which the used cotton-wool pledgets, extracted threads, tampons, drainage tubes, and used instruments are to be laid.

(*d*) A wound-syringe, or an irrigator, to syringe through the drainage tubes (those extracted, and those to be newly inserted), or with which to syringe the suppurating wound-cavities when necessary.

(*e*) The glass with drainage tubes must be always at hand, should the surgeon desire to insert fresh ones.

(*f*) Sometimes the skin under the dressing is much reddened ; at times this causes intense itching and smarting, which may be alleviated or removed by sprinkling with zinc dusting-powder (pure starch-powder and oxide of zinc, mixed in equal parts), or by applying salicylic-vaseline.

After the dressing, where possible, the patient should always be removed to another bed ; if not practicable, then at least his bed should be re-made, and he should be placed comfortably in it. Give him some beef-tea or wine, and let him remain quite undisturbed.

However complicated these proceedings may appear in print, in practice they prove much more simple than the old method of dressing. Even were the new method much more complicated it must have the preference, not only because the patient suffers very much less during the changing of the dressing itself, but because frequent dressings are much less necessary, and the average time now taken for the healing of wounds is not more than a third or a fourth of that formerly required.

Used dressings must be burnt; in hospitals, where economy is necessary, the waterproof material is taken out, cleansed with soap, disinfected, and dried so as to be usable at a subsequent dressing, but it must be again laid in antiseptic fluid shortly before re-use. After every dressing the doctors and . the nurse wash their hands; the latter cleanses the instruments, and, should they be immediately required, she lays them anew in carbolic acid solution.

As the wound progresses in healing, if ointment dressings are ordered (white-lead, zinc, boracic, nitrate of silver ointments), spread the salve with a broad knife (*never* with the finger) upon the piece of disinfected linen, shaped previously to the size of the wound.

PREPARATION OF BANDAGES FOR FRACTURES, DISLOCATIONS, AND CURVATURES.

Bandages with prepared *splints (wood, sheet-iron,* or *leather)* are now seldom used except provisionally; they can be made of the covers of boxes, cigar-boxes, bark of trees, &c., so as firmly to fix the bones of the broken limb for the transport of the injured person. These splints are applied only after the limb has been surrounded by linen, cotton, or flannel bandages; the splints are first padded with layers of linen or cotton-wool. To prevent this padding from moving out of its place, the simplest way is, to fasten it with adhesive plaster to the splints, and the latter again are secured to the limb by bandages. To give as little pain as possible, in applying and

SPLINTS.

(Provisional.)

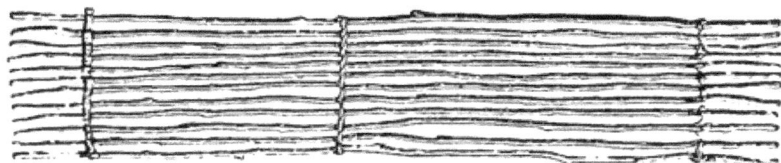

SPLINTS FORMED OF TWIGS.

STRAW SPLINT.

SPLINTS.

(Provisional.)

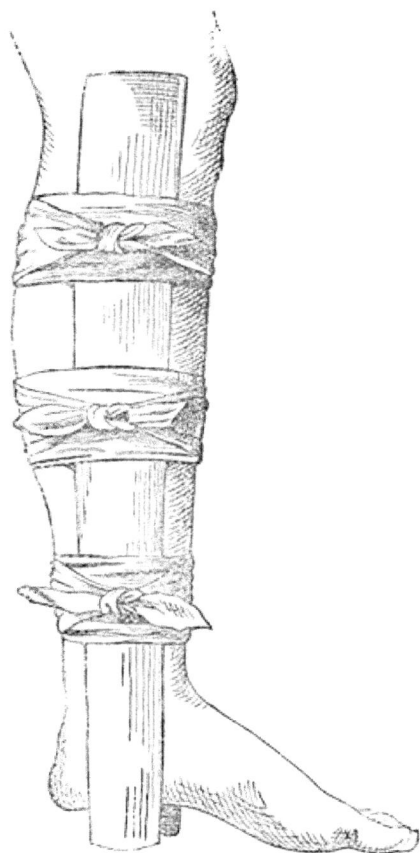

PROVISIONAL SPLINTS AND HANDKER-
CHIEF BANDAGES.

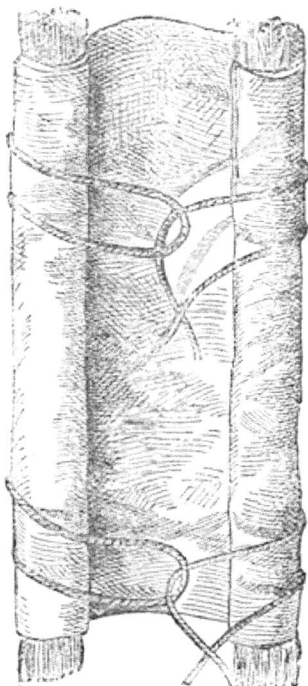

STRAW CRADLE.

removing such bandages much skilfulness, a firm, strong hand, and endurance are requisite.

As a rule, for *firmly fixing* limbs which have been broken or dislocated, or, in curvatures, limbs which have been set straight, bandages are used which, soon after their application, become firm, and are therefore intended to remain for some time as applied.

For these purposes so many methods for preparing the bandages have come into use that it is scarcely possible to test them all; it is in great part custom, fancy, or even pleasure in one's own invention that causes doctors to prefer one method of stiffening bandages to another. Albumen, starch-paste, glue, silicate of potash, cheese (curds) with lime, plaster of Paris, shellac, with bandages of the most diverse materials, strips of linen, gutta-percha, flannel, felt, and many others, have been and are still used in very varied combinations.

From my experience, the different methods of using *plaster of Paris bandages* suffice in by far the majority of cases; for lighter bandages, *starch-paste*, or a good *silicate*, especially if combined with *paste-board* or *gutta-percha* splints, may be judiciously applied.

THE PLASTER OF PARIS BANDAGE.

The finest and best plaster of Paris should be always used; it must be kept in well-closed tins in a dry place. All apparatus which have been provided at my clinic for making plaster of Paris bandages were soon laid aside by the female attendants, who returned to the older method of emplastering with the hand. This is done as follows:—from a rolled-up gauze-bandage a part is unrolled upon a table to the length of the table, and on this piece dry gypsum powder is laid moderately thick, and evenly spread by the hand, or a knife. The emplastered piece of bandage is now loosely rolled up; the plain gauze roll is further unrolled and emplastered, and so the process is repeated, until the plaster of Paris bandage is from 10 to 15 feet long. A number of such bandages may be prepared and kept in reserve in the plaster of Paris box.

When the bandage is to be applied, first lay around the limb, both at the upper and at the lower ends of the part to be bandaged, a strip of linen, dipped in water, the "binder"; then place a layer of cotton-wool around the whole part of the limb to be enclosed by the bandage.* The best way to do this is, to *cut the sheets of cotton-wool into hand-breadth strips*, and wrap them as a bandage around the limb to be emplastered. This cotton-wool must therefore be cut ready; it is fastened on by gauze-bandages (under-bandages), also kept ready. When the under-bandages are about to be applied, the plaster of Paris bandages must be laid in cold water in a deep bowl; if not too tightly rolled, they will be sufficiently saturated by the time the under-bandaging is finished. Bed and floor are kept from being soiled by the gypsum by linen or waterproof sheets spread over them. If specially rapid setting of the plaster of Paris be desired, add a tablespoonful of powdered alum to about three pints of water, in which lay the gypsum bandage; but too much alum must not be taken, nor must the bandages lie too long in the water, or the plaster of Paris will crumble. Should the bandage not be sufficiently firm, it may be strengthened by laying moist plaster of Paris over it. This paste is made as follows :—the nurse puts the

---

* When the plaster of Paris bandaging is finished, the extreme edges of this "binder" are turned over on to the bandage, thus "binding" its edges.—R. G.

requisite quantity of powdered gypsum into a bowl or hand-basin, and adds a little water; stirring continuously, she adds more water until a thin paste is made; this is spread upon the bandage with a spoon, or by the hand. All must be quickly done or the paste becomes stiff, even in the bowl, and is no longer soluble in water. Modifications of the plaster of Paris bandage are numberless. When bandaging is to be done with suitably cut pieces of linen or flannel, not dry-emplastered but only drawn through the paste, applied, and fastened by a bandage, then to know how to prepare plaster of Paris paste, with some experience of the rapidity with which it solidifies, is essential.

It is difficult to free the hands quickly from the plaster of Paris paste unless they were previously rubbed with grease, specially around the finger-nails. Rub the wet hands well with common salt, and then wash them; the nails are cleaned with a nail-scraper, and, when this is done, scrub the hands thoroughly with a small nail-brush. If several bandages have been applied one after the other, the skin of the hands does not get to its normal state, nor the nails free from the gypsum, till the next day.

The nurse must also practise *cutting open plaster of Paris bandages;* for this, strong scissors, like garden shears, have been invented, and if the bandage be not too thick, they are suitable. The simplest way is, to cut the bandage open with a *very sharp strong knife*, the shape of a garden-knife, and to cut into it obliquely, whilst a finger is pushed under the bandage. Great care must be taken not to cut the patient or one's self, and the bandaged limb must be moved as little as possible. Cutting the bandage open is facilitated if it be moistened with water along the line where it is to be cut. This knife is also the best for cutting out a hole, a "window" in the plaster of Paris bandage; this, too, must be carefully practised, and it should be cut out from two to three hours after the bandage has been applied. The edges of the " window " must be bound round with linen, fixed by plaster of Paris paste, or by collodion.

STARCH BANDAGE.

Although light plaster of Paris bandages can be made for patients able to move about in them, yet their durability is too slight, unless the gypsum is combined with other substances (glue, gum, organdy bandage). Long before the plaster of Paris was used, very light, but very firm, bandages were made with bookbinders' starch-paste. This paste is made by mixing about 5 ounces of starch-flour in a pint of cold water, and, whilst stirring, adding about $2\frac{1}{2}$ pints of boiling water.

Calico or linen bandages are best fitted for bandaging with starch-paste. Before applying, unroll them, then draw through the starch, re-roll, and lay ready for use. Starched bandages alone, even when stiff (after 36 to 48 hours), are not sufficiently resistant—they must be strengthened by splints. For this purpose rather thick pasteboard, cut into suitable splint pieces, dipped in water until soft and pliant and afterwards plastered with starch, was formerly used. When applying the bandage the strips of cotton-wool and gauze under-bandages must be used as in the plaster of Paris bandage; over them the pasteboard splints are placed, the starch bandage covering the whole. This adheres perfectly close, is very light, but dries slowly. If the fractured limb has a tendency to return to the abnormal position, strong wooden splints must be temporarily applied over the bandage, and they must not be removed for two days. More solidity will be obtained in the beginning if, instead of pasteboard, thin wooden splints of so-called "shoemaker's splinter"* are used, and, by immersion in water, these may be made more pliable. Strips of sheet iron, or several lengths of telegraph wire laid together, have been used as splints, especially in war time, when help must be obtained in any way possible.

The best splints are formed of *strips of gutta-percha*, of proper length and breadth, in combination with the starch bandage. These are cut with a strong sharp knife (plaster of Paris, or garden knife) from rolled gutta-percha, or gutta-percha straps are taken, such as are used on driving-wheels in factories. When the under bandage has been laid on the limb, the gutta-percha splints are grasped at one end by a dressing forceps and are held in nearly boiling water until soft enough to be shaped and pulled in all directions at will; the splints are then applied to the limb and fastened by the starch bandage. The warmed gutta-percha, in cooling, hardens almost more rapidly than plaster of Paris, so that starch and gutta-percha bandages would certainly have supplanted all other forms were not the gutta-percha rather expensive, and were not freshly-boiled starch-paste, and a considerable quantity of hot water for softening the gutta-percha splints, always required; the bandage with gutta-percha is rather heavier than with pasteboard. All these considerations are not of much consequence to rich hospitals and in private practice, but in practice amongst the poor, and in war, they are important. Thus, in the two last-named cases, the plaster of Paris bandage has forced all other bandage materials into the back-ground.

Silicate of Potash—*water-glass*—of more or less good quality, can be obtained of druggists in large towns. It

---

* These are thin strips of beech wood, about 1/16th of an inch thick, such as are largely used on the Continent for putting between the soles of boots.—Tr.

is dissolved in water, like starch. When bandaging, the silicate of potash solution is laid on the linen or cotton bandages with a painter's brush. If the solution is not too thin, the water-glass hardens in less time by some hours than the starch bandage, and has the advantage of being very light ; without stiffening splints, however, it affords no more resistance to displacement of the bones than does the starch bandage alone. This form of bandage often fails, from the bad quality of the silicate.

If parts of the body are to be protected against injury from contact, or from pressure of any kind, then a bandage of three to four layers of organdy is sufficient.

Although it is not the proper duty of a nurse to apply bandages to fractured bones, because she neither can nor ought to take upon herself the responsibility of the right position of the extremities of the broken bones, and their union in this position, yet she must acquire a certain degree of *practice in applying bandages*, as occasionally it may be necessary for her to apply a bandage. She must be able to apply a roller-bandage faultlessly from the hand to the shoulder, from the foot to the hip ; to bandage the head alone, then the head and throat, then the chest, the shoulder, the groin, with close-fitting, but not too constrictive, turns of the bandage. A cotton, a linen, or an elastic bandage can be applied more or less tightly. One must learn by practice to know exactly the degree of pliability of the material used, as well as the suppleness of different parts of the body, and thus be able to determine beforehand how tightly the bandage should be applied.

It is also very important that the nurse should know how, properly, to use triangular and square handkerchief-bandages ; these are more simple and more quickly applied in many dressings than ordinary bandages. Here I would direct special attention to the method of applying the arm hand-kerchief-bandage *(mitella *)* ; it is not always easy so to put it on that the arm shall be firmly supported and the weight shall not be too heavy on the neck.

The accompanying illustrations represent a variety of bandagings with roller, and with handkerchief-bandages ; how to apply them is taught in the Practical Courses of Instruction, and such modes of bandaging should be thoroughly practised by nurses.

The materials for practising are stout linen bandages (bandages of soft texture are easier to apply, but their flexibility disguises the faults that are made), of different widths, and handkerchief-bandages of thin cotton or linen.

The four-cornered handkerchief-bandage is about 36 inches to 38 inches square; if this be cut in two diagonally, from corner to corner it makes two large triangular cloths ;

and if one of these be again halved , two small three-cornered pieces is the result.

The appliances (sandbags, grooved splints, Petit's boots,† slings) for fixing the position of injured limbs must be so well known to the nurse that she can fetch and arrange them for use, and be able to correct any small defect in them.

If fractures or operations on bones are accompanied by wounds requiring antiseptic dressing, the nurse must make every necessary preparation.

* Mitella : a triangular bandage.—Tr.

† Petit's boot is a grooved splint for the leg with a grooved support for the sole of the foot, and with a hole to receive the heel.—R. G.

TRANSPORT OF THE SICK.

To accompany the sick or wounded from their homes to the hospital, or from the hospital to their homes, is a frequent duty of nurses. When so accompanying the nurse must first of all direct her attention to the careful lifting of the patient from the bed, then to the carrying down stairs, and then to placing him in the carriage, or the railway-carriage.

It is, therefore, as useful as it is necessary that she should know the use, and be instructed in the arrangements, of the different appliances and means of transport in vogue in modern times.

HANDKERCHIEF BANDAGES.

A FALSE KNOT.

THREE-CORNERED BANDAGE
FOR THE HEAD.

TRUE, OR SAILOR'S KNOT.

BANDAGE FOR THE EYE.

(*First Stage.*)   (*Complete.*)
FOUR-CORNERED BANDAGE FOR THE HEAD.

L

HANDKERCHIEF BANDAGES.

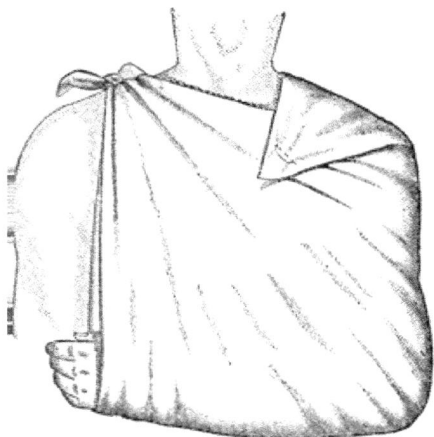

FOUR-CORNERED BANDAGE
FOR THE ARM.

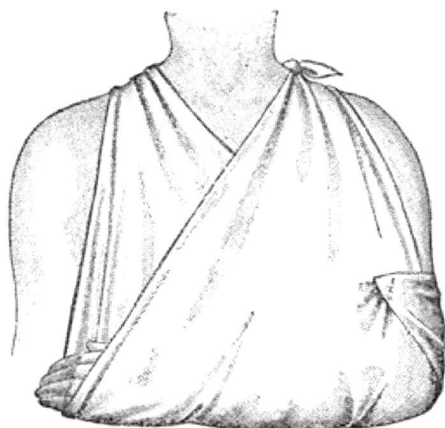

THREE-CORNERED BANDAGE
FOR THE ARM.

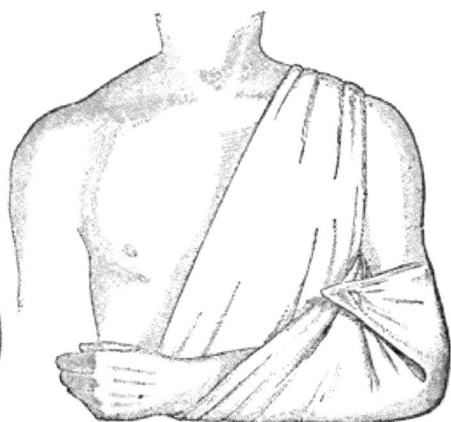

THREE-CORNERED BANDAGES FOR THE ARM.

HANDKERCHIEF BANDAGES.

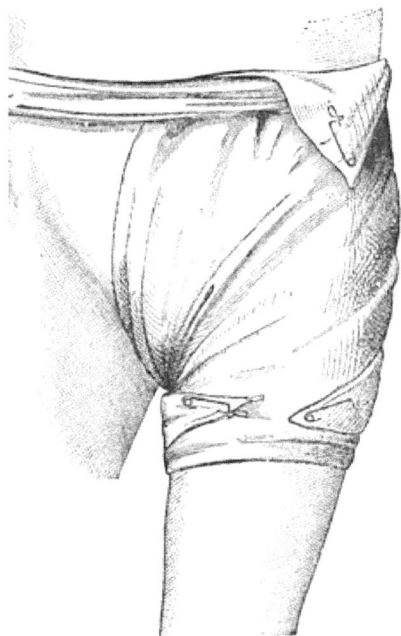

E-CORNERED BANDAGE FOR THE HIP.

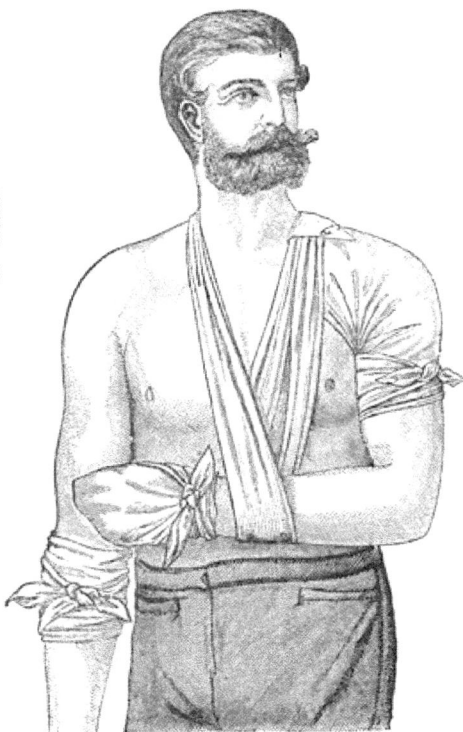

BANDAGES FOR SHOULDER, HAND, AND ELBOW;
SLING FOR THE ARM.

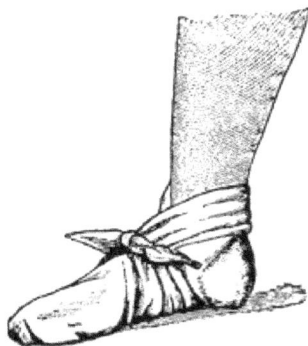

HANDKERCHIEF BANDAGE
FOR THE FOOT.

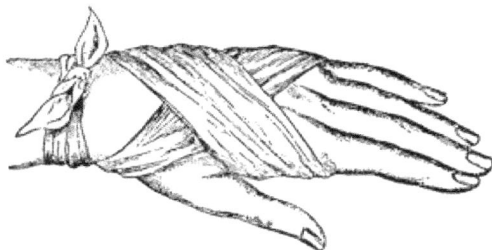

"CROSS" BANDAGE FOR THE HAND,

HANDKERCHIEF BANDAGES.

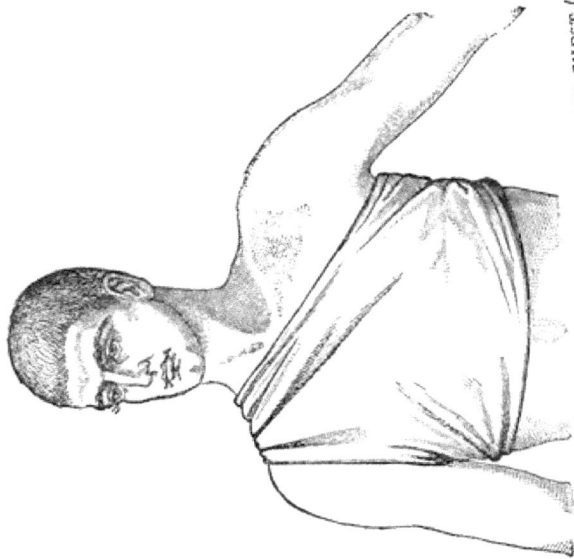

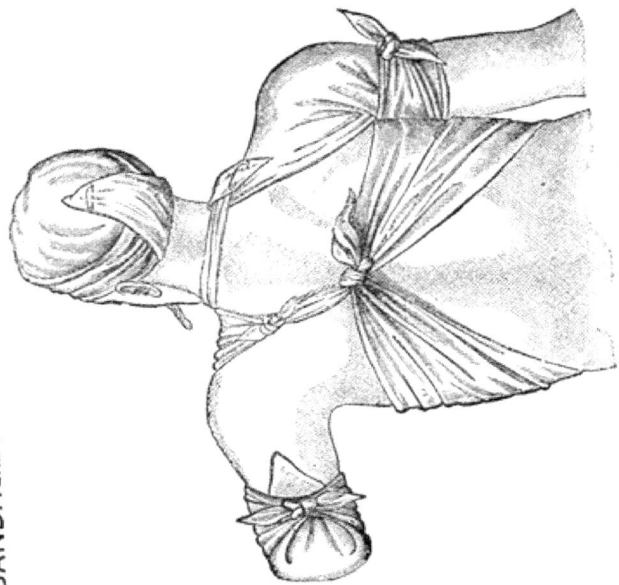

THREE-CORNERED BANDAGES, FOR CHEST (*front and back view*), SHOULDER, ARM, AND HEAD

DIVIDED BANDAGES.

DIVIDED BANDAGE.

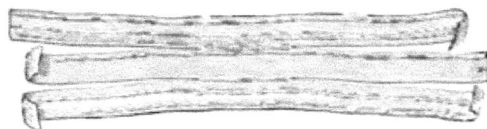

DIVIDED BANDAGE (SIX-TAILED).

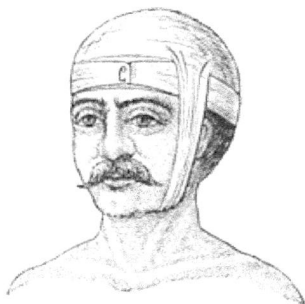

SIX-TAILED BANDAGE FOR
THE HEAD.

DIVIDED BANDAGE DRESSINGS FOR THE HEAD.

DIVIDED BANDAGE FOR THE JAW.

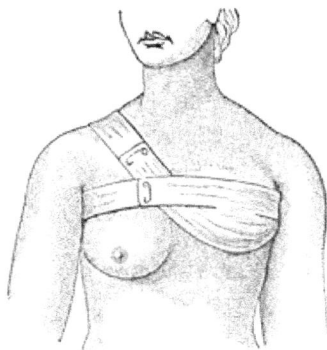

DIVIDED BANDAGE FOR THE BREAST.

T-BANDAGES.

 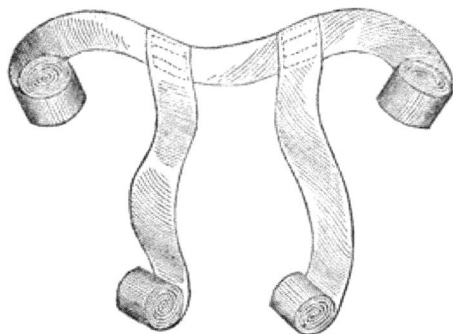

T-BANDAGES.

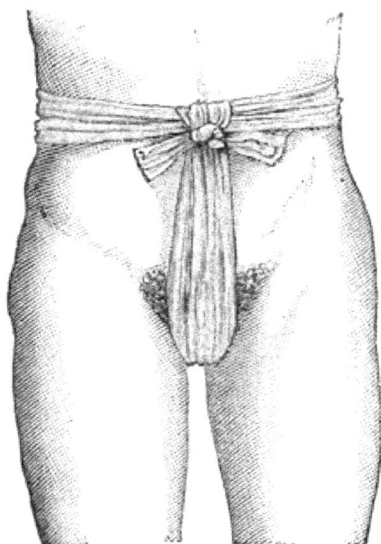 

T-BANDAGE.   CHEST BELT (DOUBLE T-BANDAGE).

DRESSINGS AND ROLLER BANDAGES.

DRESSING FOR BREAST OF COMPRESS AND
BANDAGES (APRON DRESSING).

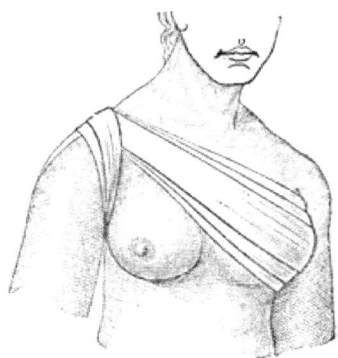

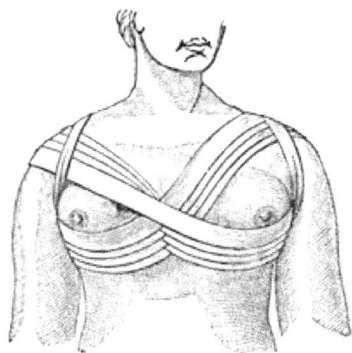

HANDKERCHIEF BANDAGES FOR JAW.

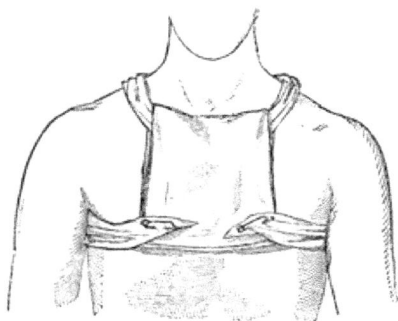

APRON DRESSING.

SUSPENSORY BANDAGES FOR BREAST.

ROLLER BANDAGES.

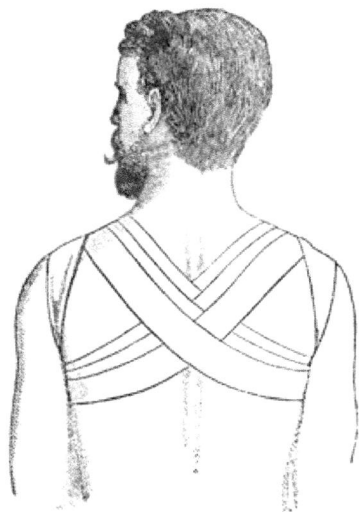

BANDAGE FOR THE EYE.

"CROSS" BANDAGE FOR THE BACK.

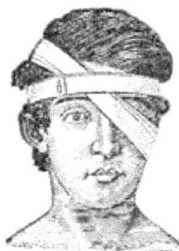

HALTER BANDAGE.

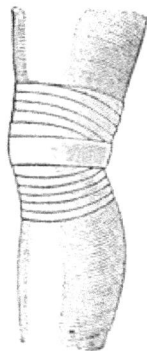

FOR KNEE OR ELBOW.

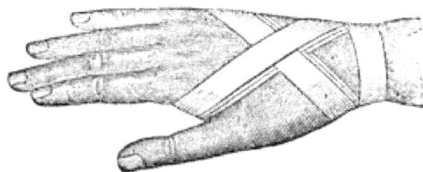

"CROSS" OR FIGURE OF 8 BANDAGE.

ROLLER BANDAGES.

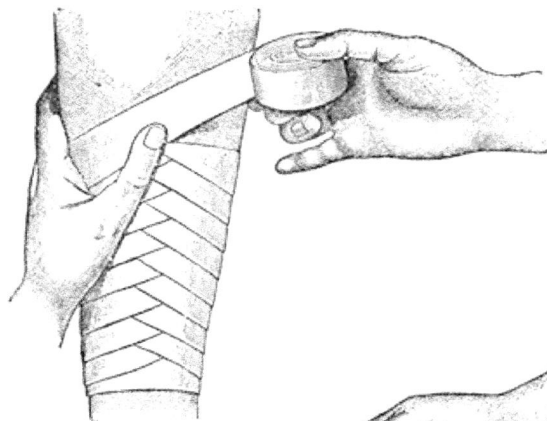

No. I.

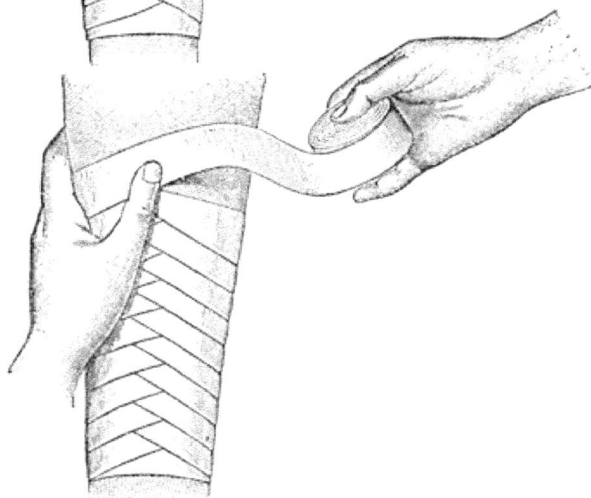

No. II.

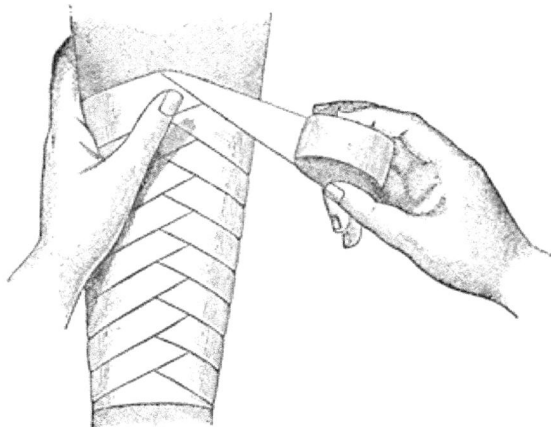

No. III.

BANDAGING: "TURNING" THE BANDAGE.

M

ROLLER BANDAGES.

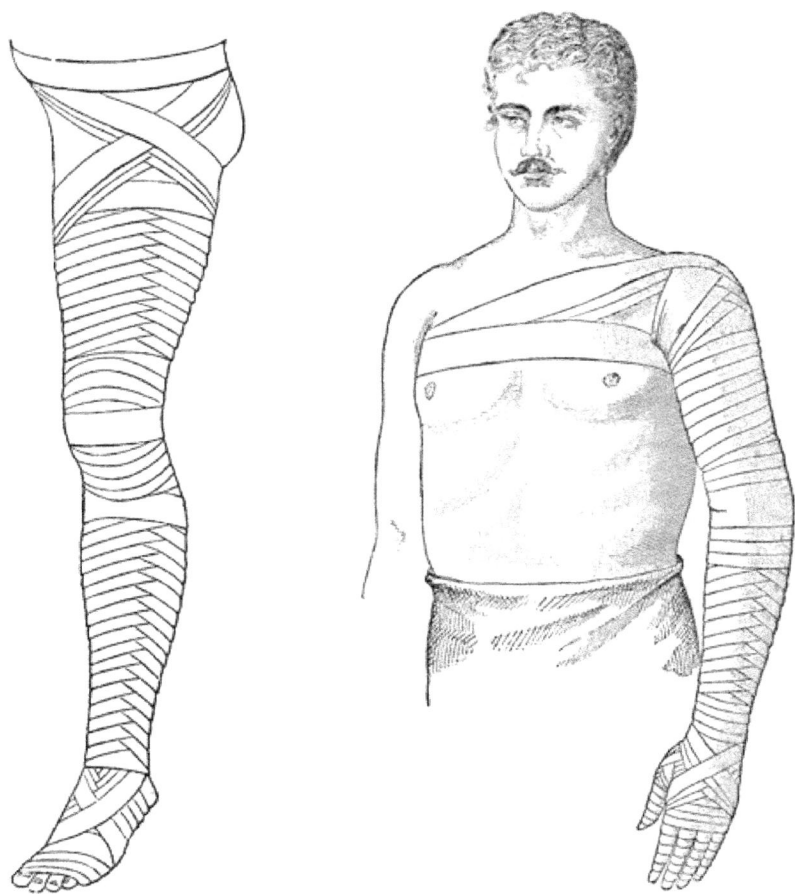

LEG, ARM, AND HAND, ROLLER-BANDAGED.

ROLLER BANDAGES.

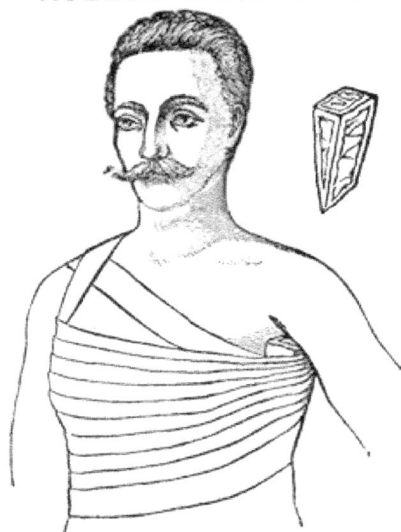

I.

II.

DESAULT'S BANDAGE.

III.

CHAPTER V.

OBSERVATION AND CARE OF FEVER PATIENTS GENERALLY.

FEVER. ACUTE, CHRONIC DISEASES. RAISED TEMPERATURE THE CHIEF SYMPTOM OF FEVER. MEASURING THE TEMPERATURE OF THE BODY BY THE THERMOMETER. FEVER CURVES. COUNTING THE PULSE AND THE RESPIRATION. FEVER DELIRIA. DUTY OF THE NURSE.

FEVER (from the Latin *febris*, and this from *fervere*, " to be hot,") is a condition occurring in many diseases, and is, perhaps, always the result of the entrance of the smallest morbific organisms into the body and its struggle against them. It is extremely rare (except in injuries to certain parts of the spinal cord ; after blood-letting ; certain treatments, such as the dry-food treatment) for fever to arise from other causes. A disease is inflammatory when the patient is almost always in a heat (in fever), but such illness usually does not continue for weeks ; it runs its course violently and quickly, is " acute," as opposed to other diseases lasting for months or years, the course of which we designate " chronic " ; the distinction is not always quite rigid : illness, which originally appeared as acute, can become chronic, and that which appeared as chronic can become acute. Although acute diseases without fever are extremely rare, it frequently happens that chronic patients are violently feverish, especially in the evening. To observe these conditions is an important duty of the nurse, and she must conscientiously report her observations to the doctor; it is best—as, in fact, with all her observations—to write down the incidents that happen in the illness, with the date and hour of the same. The fever can vary very much in degree—the temperature of the body is a test of it, as is also the frequency of the heart and pulse-beats.

The heat which the body inherently possesses is termed specific heat, or blood-heat; it arises from the chemical processes at work in the body by means of respiration, circulation, and muscular action. We make a distinction between the specific heat of the body and that casually, or intentionally, imparted to its surface by hot surroundings (hot water, hot compresses, hot air). This temperature of the surface of the body, depending upon external influences, is of no importance to the question whether the patient is feverish or not. Man belongs to those organisms that produce much warmth and only relinquish it with difficulty—that remain equally warm internally (warm-blooded), and therefore can live in all zones, hot or cold. All warm-blooded creatures are not equal in temperature, the mouse, 105·8° F. (= 41° C.), and the swallow, 111·2° F. (= 44° C.), have the highest temperatures, the dolphin, 95° F. (= 35° C.), the lowest. The temperature of the cold-blooded is adapted to the external world in which they live; hibernators have a temperature of 41° F. (= 5° C.) in winter, and only perish at 32° F. (= 0° C.), whilst man, if compelled, by his external circumstances, to yield up so much of his *internal* temperature as to reduce *it* for any length of time below 86° F. (= 30° C.), cannot continue to exist.

To measure the temperature of the more deep-seated parts of the body and of the blood, the mercury end (the best form of the end is cylindrical and not spherical) of a *thermometer* must be introduced either into a cavity of the body (mouth, anus), or, closing the armpit, or the groin, the thermometer is laid into the hollow thus formed. For this, thermometers are used with the scale of degrees sub-divided into tenths, according to Celsius. Were such a thermometer to shew, like ordinary thermometers, all the degrees from the boiling point of water to its freezing point, it would be much too long. Accordingly, the index of degrees (scale) of this thermometer is marked with no more degrees than are necessary to indicate the lowest, 29° C. (= 84·2° F.), and the highest, 42-43° C. (= 107·6° to 109·4° F.), blood-temperatures of living men. A thermometer should be chosen with not too fine a scale, or one with a convex, polished cylindrical glass which magnifies the column of mercury, so that, even with a bad light, it can be read off accurately. At first the nurse will have some trouble to see exactly the extremity of the mercury, but she will soon learn by daily practice. For such practice she should never inconvenience a patient, but should frequently hold the bulb of the thermometer in her closed hand —then she will see the mercury rise, and will observe that it sinks slowly again when she relinquishes the lower, and holds the instrument by its upper, end.

The temperature in the armpit is thus taken : the thermometer is placed in the bare armpit (the mercury-end right up in the armpit), and this is closed by the patient laying his arm, bent at the elbow, diagonally across his chest ; in so doing care must be taken that, whilst the thermometer protrudes from the armpit in front, it does not drop down at the back. As it is not comfortable to keep the arm long in this position, place a pillow under the elbow as a support, or lay the patient somewhat on his other side. At first the mercury rises rapidly, then more slowly; generally, in from ten to fifteen minutes, the maximum temperature is reached.

Before noting the temperature it is a rule to leave the thermometer in the armpit until the mercury has remained at the same height for some minutes, or to leave it there each time for ten minutes, and thus avoid disturbing the patient by too frequently inspecting it. Most patients are very little inconvenienced by lying quiet for a while whilst the temperature is being taken, but some are restless and impatient. For them a method is adopted by which the necessary time is shortened. If the arm is kept close to the body for some time before putting in the thermometer, the temperature of the armpit is more constant, and the thermometer is left in for about five minutes.

Where no one capable of reading the temperature is at hand, thermometers are made called maximum thermometers, in which the top part of the column of mercury keeps at the highest point to which it rises, even when the instrument is afterwards placed in a lower temperature. After lying in the armpit from five to ten minutes it is removed, and at his next visit the doctor himself will note the temperature. Before it is again used, the column of mercury must be shaken together by rapid, downward, jerky movement.

As thermometers do not always accurately agree one with another, the rule is, to use one instrument always for the same patient.

By frequent examinations of healthy persons it is found that the specific heat of the same person fluctuates about one degree Celsius in the course of twenty-four hours; but even the specific heat of healthy persons differs—the highest temperature in health (from 4 to 5 o'clock p.m.) may reach 99·5° F. (= 37·5° C.), and the lowest (at night, towards the morning) may sink to 97·25° F. (= 36·2° C.). A temperature in the morning of 99·5° F. is to be regarded as abnormally increased, whilst the same in the afternoon is taken for normal. By very violent movement the blood-temperature can be often somewhat raised, but only temporarily ; in testing temperature at the bedside, this would not be taken into account unless the patient had had severe convulsions shortly before.

In disease, the temperature may rise to 107·6° F. (= 42° C.), and may fall as low as 95° F. (= 35° C.). The height of the specific heat is the measure of the height of the fever, and high fever is always unwelcome in disease ; but occurring once, or a few times only, it does not always indicate danger, as many persons believe. It is equally maintained high fever, continuing for days, that is evidence of the severity of a process of disease. For the recognition of the process of the disease (diagnosis), and for the foretelling of its course (prognosis) so as always to estimate the symptoms correctly, much experience is necessary. The nurse must guard against communicating to the patient, or to those around him, any of her fears of threatening danger—that is the duty of the doctor.

For taking temperature in the rectum the patient must lie on his side, drawn up together so that the lower oiled end of the thermometer can be introduced like an enema ; in so doing with resisting, delirious patients, or with refractory children, care must be taken that the thermometer is not broken. The temperature in the rectum is nearly one degree higher than in the armpit ; consequently, it must be specially noted whether the temperature was taken in the anus or in the armpit, and the conclusions must be drawn from the results of the measurements in one only of the localities named.

If the doctor considers it necessary for the temperature to be taken more than twice daily, he will state the hours of the day or night when these observations must be made.

Easily to represent the course of the fever by an ascending and descending line, forms are provided, on which the thermometer scale is drawn in continuous horizontal lines; these are crossed by perpendicular lines; the equal squares thus formed indicate a day of illness (that is, the hours of a day of illness). On the temperature being read off it is at once marked on the form by a dot, and by connecting several dots a line is made, termed the *fever-curve*. When no such forms are to hand, the temperature is noted upon a sheet of paper (which must be carefully kept) so that the temperatures taken at the same hours daily stand in line. The nurse can make her remarks opposite the respective hours, noting special occurrences, giving of medicines, &c. For these things she must rely upon her notes, and not upon her memory.

The reason why measurements of temperature have found such general acceptance as tests for determining the height of fever lies in the fact, that to take the temperature is far more easy for the laity than to count the beatings of the pulse, because the heat of the body is not influenced by affections of the mind, as is the case with the action of the heart, and, consequently, with the pulse to a high degree. *Feeling and counting the pulse*, however, have not become superfluous, because of the measurements of temperature. The doctor often draws highly important deductions from the frequency and kind of pulse—indeed, they indicate to him a continuously approaching danger much more directly than does the temperature. But as this always takes place with due consideration of the whole aspect of the disease and its previous course,—and the quality of the pulse as an isolated symptom is scarcely ever of decisive significance,—so an unlearned person would err much more readily if he were to draw conclusions from this symptom than he would from the specific heat. The doctor welcomes the report of the frequency of the pulse as observed during his absence, and this a trained nurse should understand, and she should also learn to note other important symptoms indicated by the pulse. We are enabled to determine, from the *volume* of the pulse, whether the heart is pumping much or little blood into the vessel; whilst *rapidity* indicates whether the heart is contracting quickly or slowly; and *resistance to pressure by the artery* indicates the *force* of the heart. A small, soft pulse (in convalescents, for instance) demands great caution when the patient sits up, or is transferred from one bed to another. It is well known that the pulse is usually counted somewhat above the wrist, on the thumb side, where the radial artery lies near the surface. How best to do this, and how to guard against mistakes in the doing, can only be learnt by demonstration. The number of pulse-beats in a healthy adult fluctuates from 60 to 80 per minute. Children have more—newly-born infants, 120 to 140; old people, less; lively persons more than phlegmatic; any excitement, even the presence of the doctor, an examination of the body, a dressing, pain, etc., may have the effect of increasing the pulse-beats by from 20 to 30; the number can be so increased, and the pulse may be so difficult to feel (so *small*), that exact counting becomes impossible. In such a case, rather let the nurse state that she can no longer count exactly than make definite, but incorrect, statements to the doctor. In most feverish diseases, the frequency of the pulse-beats *(pulse-frequency)* stands in definite relation to the height of the temperature, but there are many exceptions; the combination of a very low temperature with great rapidity of pulse in severe long-continued disease is not rare.

If a person is very hot, and his heart and his arteries beat very rapidly, then, proportionally, he usually *breathes more frequently.* A healthy, quietly-recumbent adult breathes eighteen respirations per minute; the new-born infant, from forty to seventy; in fever the number is increased, but when the respiratory organs, particularly the lungs, are affected, there is great increase of difficulty in breathing, and the number of respirations taken alone is no longer a correct test of the fever. The number of respirations is thus determined : let one hand rest lightly on the chest of the recumbent patient, hold the watch with the other, and count how often the chest heaves during one minute. This must be done several minutes in succession, for at first, knowing he is observed, he is likely to breathe too hurriedly, or to retain his breath. Whilst in quiet, almost inaudible breathing, no great muscular effort is necessary, and only the anterior abdominal surface rises at the respiration, so difficulty in breathing will declare itself by very violent movements of the chest and nostrils, and the swelling of the muscles of the neck and chest. In certain forms of difficulty in breathing (croup in children), the concavities of the neck are seen drawn deeply in, and a loud respiratory sound is heard. Respiratory movements, which—like the pulse—can be sometimes irregular, are best observed (especially with children) in patients when asleep. Therefore a nurse should never, on any account, wake a sleeping sick child when the doctor comes.

Frequency of pulse and of breathing can also be represented by *curves*, if suitable forms be supplied; except for scientific purposes, these forms are little used.

Poets and novelists often speak of " feverish phantasies," of "feverish dreams," of the "delirium of fever." It is true that, in persons suffering from violent fever, hallucinations of various kinds often occur, but it is equally true that these are frequently wanting with the highest fever-temperatures observed. These fever deliria are mostly consequent upon brain-irritation, resulting more from a rapid exchange of hyperæmia * and anæmia † of the brain and special abnormal admixtures in the blood (blood-poisoning), than from very high temperature of the blood.

At all events, the observation is perfectly accurate that feverish patients are, bodily and mentally, essentially irritable, and everything that excites them, such as dazzling light, noise, exercise of their intellectual powers, &c., must be avoided. Such patients are very sensitive to variations in temperature. If a fever rises rapidly—*i.e.,* the temperature is increased some degrees in half-an-hour or less—then the patient feels as if everything about him were very cold—he begins to feel chilly, he shivers, his teeth chatter, he shakes *(rigor, chill)*, and this feeling increases when he moves, or when one uncovers him a little. If, at this period of the fever, the specific heat be measured it will be found not very low, as the patient thinks, but very high. After this has continued for a while (from a few minutes to half-an-hour), *a period of dry, burning heat* succeeds—the specific heat is very high, mostly highest at the commencement of the stage of dry heat ; but the perceptions of the patient have become more correct—he realises that the room has not become colder, but that he has become hotter. Rigors should always be noted in writing by the nurse, with date, hour, and duration of the same.

Should the specific heat rise slowly, the patient does not feel a chill, but he experiences a gradually increasing feeling of heat, frequently accompanied by dragging pains in the loins and back.

* Hyperæmia : " Excess of blood in a part."—MAYNE.
† Anæmia : " Deficiency, or poorness of blood."—MAYNE.

In a regular attack of fever, as is most clearly seen in " intermittent fever," or "ague," *the stage of perspiration* follows that of dry heat, and may continue for an hour or more. The temperature often sinks rapidly during this stage, and the patient is much relieved.

In severe, long-continued feverish illnesses, the dry heat continues without being followed by perspiration, and only towards the end of the illness, in the transition towards recovery, does perspiration come with increasing health. In patients seriously ill the outbreak of perspiration does not always denote recovery, as the dying can be covered with perspiration.

Now, as far as *hallucinations* (deliria) are concerned which are combined with the fever heat, these are generally dream-like pictures, such as occur to healthy but easily excitable people, particularly when half asleep ; the patients talk in a low voice to themselves, and make movements like one in a dream ; they are easy to waken from these dreams ; when spoken to they are perturbed and quickly come to clear consciousness, but soon return to their former condition. These low, fever deliria are tolerably frequent. Those in violent fever seldom appear fully awake, and their hallucinations, that is to say, the pictures of their diseased imagination, are so real that they see persons before them who are not present, hear them speak, dispute with them, believe that they are attacked by them, defend themselves, spring out of bed, try to run away. They suffer from *fever delirium*, which resembles that from alcohol or chloroform, and they cannot be freed therefrom until the fever abates.

It is often very difficult to keep such patients in bed, as the strength which even weak persons then evince is very considerable. Great lassitude almost always follows high fever, and is accompanied for a time with a feeling of dulness in the head, like that which a healthy person suffers after heavy dreams.

One phenomenon is always combined with fever, viz., *great thirst*. Not only from increased difficulty in breathing do fever-patients lie with the mouth open, but, from the raised specific heat, the mouth dries more rapidly, and the patients require to drink freely. They mostly desire cold water or ice; but if the feverish illness continues long, a change of beverage is always welcome.

In treating fever patients the nurse may yield to their wishes for warmth by laying on several blankets during the period of the chill; and later, by cooling the room during the stage of heat, but not too much, for, after perspiring, the room easily becomes too cool for him. When the head is hot, a cold compress may be laid upon the forehead without injury. If severe perspiration occurs, the nurse waits until it is over —then she changes the linen, rapidly dries the patient (naturally with the room well warmed), and removes him to another bed. That illness can be produced by the "checking of perspiration " may be an error; it is much more likely that an acute new malady, or a complication, has arrested the perspiration; still, patients when perspiring should never be needlessly uncovered, for it is generally very unpleasant to them, and decline of the fever-heat may be arrested. It is neither necessary nor judicious to rouse a feverish patient from a state of light delirium. Should he become very restless, and suffer from violent delirium, with raving, the nurse must obtain help to prevent him from jumping out of bed, or out of window.

Both in feverish and in feverless illnesses many duties fall to the nurse; how to discharge them, she must learn by fre-

N

quent practice.   Among such duties are the following : giving
hypodermic injections, applying mustard plasters, cataplasms,
leeches, blisters, ointments, plasters, suppositories, cupping-
glasses ; injections into the rectum, the nose, the auditory
passage, laying compresses on the eye, instillations into the
eye, insufflating powder ditto, catheterism, administering
drops, powders, or pills, vinegar fumigations, the management
of inhalation-apparatus, &c.   Every duty should be done
quickly, without more disturbance to the patient than necessary.   Every apparatus, every syringe must be faultlessly
clean before use, and must be thoroughly disinfected.

CHAPTER VI.

NURSING IN EPIDEMICS AND IN INFECTIOUS DISEASES. PRECAUTIONS AGAINST INFECTION. DISINFECTION.

OF the Epidemics which arise, and may recur, *Typhoid fever (nervous fever)*, diarrhœa *(dysentery)*, and possibly also, *epidemic cerebro-spinal meningitis*, like *cholera*, brought to us from time to time from Asia, are not contagious from person to person ; nevertheless, the morbid matter is developed in the secretions under certain frequently occurring conditions, mainly in evacuations from the bowels, and, passing from these into healthy persons, may produce the disease. Hence how vastly important it is that the evacuations from the bowels shall be adequately disinfected : omitting to do this may bring danger to a whole community.

Typhus and *relapsing typhus* (both fevers very rare in Austria), *small-pox, measles, scarlet fever, whooping cough, mumps, erysipelas,* and *diphtheria* are contagious from person to person. The contagion is spread, in rare cases, by persons who carry it in their clothes, but who themselves are, and who continue, quite well.

Malarial fevers are not communicable by contact, or by secretions. They are never, in any form, infectious; they can *only* be contracted where they arise from the earth. Other contagious diseases (hydrophobia, syphilis, anthrax, glanders), which always depend upon special and direct transmission, we shall not consider.

TYPHOID FEVER.

Enteric fever, termed briefly *typhoid,* or *nervous, fever,* is spread all over Europe, and was formerly the most frequent of all epidemics. Man receives it as he does all the other before-named diseases—always from without, and it is produced by infinitely fine seeds *(spores)* of little plants, most minute fungi, only to be perceived by the aid of powerful magnifying glasses. These spores enter the blood and, for a time, germinate and grow exuberantly in the blood itself, and in the tissues of the body (also in the lower part of the intestine, where they cause ulceration). In so doing they withdraw, not only material constituents of the blood and of the tissues in order to live, but, in their growth, they produce in themselves elements, which, like the juices of poisonous plants, operate to the highest degree injuriously on the blood. If the patient do not die from poisoning by the germination and growth of the fungus-spores, then, sooner or later, the fungi will perish and the patient will recover, unless the destruction effected in the different organs by the fungi be not of itself the cause of death, which, in typhoid unfortunately, frequently happens.

Researches of late years have proved, with tolerable certainty, that all these epidemics and contagious diseases arise and pass away as described; that, therefore, for each of these and for many other diseases, fungus-germs with specific properties exist. Man is only occasionally one of the hosts for these minute organisms. From men living together in houses, villages, and towns, these diseases spread until they become pestilences or epidemics.

Of the nature of typhoid, and the way in which it is produced and disseminated, the following is known : the typhoid-producing fungus seems to exist almost everywhere in the upper earth-strata of Europe, and to be able to propagate itself very easily. Man receives it in the dust he inhales and swallows, very frequently in the water he drinks, or in the vegetables he eats. It is impossible to detect typhoid-poison in the air, the water, or the earth by its scent. No disease-virus (fungus-spore) mentioned is perceptible by its smell ; it is only visible when it is most powerfully magnified, and even then one kind cannot be distinguished from another, or from the ordinary fungi which cause putrefaction. The absorption of disease (infection) from the earth no one can avoid ; it is a power in nature which, like many others, tends to make life uncertain. People in general cannot guard against this kind of infection—they have no water but that supplied by springs, which they must drink, even when they fear that the draught may bring disease or death. Every town cannot be supplied by waterworks with fresh spring water from the mountains ; but in this direction the State should do its utmost to protect the people. In this age, in which only settled nations and peoples are tolerated in Europe, a whole race, or a tribe, cannot abandon an unhealthy country to re-establish themselves in another locality. The number of persons able to leave, even temporarily, the places in which epidemics are developed, is always small.

One source of further propagation and conveyance of typhoid may be annihilated : the virus of the disease, which is produced in the sick person himself and is evacuated from the intestine, may be destroyed. It passes into the water-closets with the evacuations ; thence the poison rises with the offensive gases into the houses, or it dries, and is scattered like dust into the air, or it penetrates the earth and thence enters the wells. Thus man himself produces new typhoid centres. How to avoid this see pp. 201, 231.

As a rule, the typhoid poison develops in from two to three weeks *(period of incubation)* after its reception into the human body: illness then begins. In each transmissible disease referred to, the length of time varies between the reception of the poisonous germ and the outbreak of the disease, because the different germs have their different conditions of life and germination. This period, mostly passing without sign of illness, is the period of incubation, from the Latin *incumbere*, to brood, to lie upon something. To wit, in former times the sick were brought to the temples, there laid, were subjected to various ceremonies, and finally, received advice from the priests (the first doctors).

If only a small quantity of poison, operating feebly, be absorbed, or if the person attacked presents an unfavourable soil for its development (the bodies of different people act towards the poison exactly as different sorts of earth do towards different genera of plants), then, either no sickness, or one which is very slight, results : such cases doctors often designate, *gastric fever.*

In the full development of typhoid the disease lasts from three to four weeks (the "week" not reckoned strictly by the calendar, but sometimes comprising four or five days only, at others from six to ten days), but it may be protracted long beyond this by certain grave contingencies (inflammation of the lungs, brain, kidneys, salivary glands, gangrenous decubitus), which must also have time to run their course. The period of recovery is often prolonged; relapses (from poison remaining in the system and germinating only at a later period) are not rare.

THE DAILY DUTIES OF NURSES WITH TYPHOID PATIENTS.

These daily duties are not to be regarded as limited to the nursing of typhoid patients only, but as, among all acute febrile diseases, typhoid usually lasts the longest and mostly runs its course with considerable uniformity, so the duties to be discharged in nursing feverish patients can be best illustrated by it.

Everything previously said in Chapters I. and II. upon the arrangements of the sick-room, and the care of patients long confined to their beds, must here be practised.

For a typhoid patient there should always be *two beds*, and when possible, *two rooms* at disposal, in order to place him in the best attainable position, and to secure proper ventilation. From this it follows, that it is only seeming hardheartedness when doctors insist that poor patients, unable to provide two beds and two rooms, shall be treated in hospitals.

The typhoid patient should always *lie in bed*, from the commencement of the illness onwards, and should refrain

from all mental and physical occupation, all useless talking.
Typhoid often produces such slight symptoms that it is
difficult to keep the patient in such a state of rest *(typhus-
ambulatorius,* but even this can end fatally). He must not sit
up for stool, or for urinary evacuations, even if his powers be
equal to the exertion. As a rule, one person only should be
in the room, who should have charge of the nursing.

The *room-temperature* must be regulated in winter to
59° to 63·5° F. (= 12 to 14° R. = 15 to 17½° C); in effecting this,
one window must be kept open day and night. Daylight
must be softened by blinds, the patient lying so as not to look
towards the window; quietness in the neighbourhood of the
sick-room must be provided for. If several persons in one
family suffer from typhoid, they should never lie more than
two in a room, even in spacious rooms in a private house.
The long hair of female patients must always be plaited.

Keeping the patient clean is of essential importance—the
mouth in particular must be repeatedly washed out daily; if
conscious, the patient must be made to rinse his mouth
frequently, specially after taking food. Washing out of the
mouth, cleaning the tongue, and removal of mucus from the
teeth are best effected with a small piece of linen, wrapped
round the finger and dipped into glycerine diluted with an
equal quantity of water. Daily also, face, hands, neck, nose,
ears, buttocks, arm-pits, and groins must be most carefully
washed with tepid water—if the doctor permits, a tepid cleansing bath to be often given. The cold bath treatment is now
deemed beneficial for many typhoid patients. At certain
periods of the disease, fæces and urine may pass involuntarily
from typhoid patients—then frequent washing and change
of bed are highly necessary; large pieces of waterproof
sheeting must be laid beneath the sheet and underlay to

protect the mattress. If the patient can be taken in his clean bed into the adjoining room, so that, after the soiled bed-linen has been removed, the first room can be thoroughly ventilated, that is far better than fumigations, of which only those with vinegar are admissible. Besides, in a room with a window always open the bad odour is quickly removed ; but it is got rid of with difficulty where want of common sense, or exaggerated anxiety, condemns a patient to breathe impure air, and thus to aggravate the illness from which he is suffering.

It is of the utmost importance to disinfect the evacuations of typhoid patients, and at once to remove them from the room, for it is chiefly from this source that the disease is spread. Immediately before use a strong solution of sulphate of iron must be put into the empty bed-pan ; when used, and the patient cleaned and re-arranged, pour into it about $3\frac{1}{2}$ fluid ounces of crude hydrochloric acid. With these disinfectants (from smallness of cost, much used) great care must be taken not to spill any upon the linen, as stains, very difficult to remove, and holes are the result. Some doctors allow only chloride of lime to be put into the bed-pan, before and after use. Strong solutions of permanganate of potash operate destructively upon the infectious matter, and at the same time most quickly take away the smell ; should any of the liquid be spilt upon the linen, brown stains are made, which can be removed by hyposulphite of soda and hydrochloric acid (see "Sponge cleansing" page 140). All being well mixed by pouring water upon the evacuations in the bed-pan, the morbid matters which propagate the disease are thus destroyed, and it will cause no injury to empty the bed-pan into the water-closet.

It is yet more effectual to pour the evacuations into a pit partly filled with lime, and every time to cover them with earth.

This immediate destruction of the evacuations must always be carried out except when the doctor orders them to be kept unaltered; then the cover of the bed-pan must be closed securely, and the bed-pan placed in the water-closet.   After he has there inspected the evacuations, disinfecting must immediately follow.   The nurse must always observe that which is evacuated, so as to report upon it to the doctor; she must particularly notice whether blood, shreds of mucus, or such-like matters were mixed with the fæces.   In doubtful cases, the bacteriological examination of the evacuated matter will aid the diagnosis : from this the importance of such matter being kept for observation will be recognised.   Notwithstanding the greatest experience and accuracy of doctors, typhoid cannot always be distinguished from other diseases (for instance, from acute tuberculosis embracing all organs, pyæmic processes, &c.).   The bed-pan, being washed, must be further rinsed with three per cent. solution of carbolic acid.

Linen soiled by patients must be at once put into a vessel with lye,* and there left till washed.

Great consideration must be given to *the position of the patient.*   When no difficulty in breathing exists, the typhoid patient may lie tolerably level; if in a higher position (unavoidable, when concurrent with lung disease), he slips down in the bed easily, his head falls upon his breast, his body is doubled up, and respiration is much impeded.   To frequently raise him, and to induce him to breathe deeply, become necessary.   With insensible patients, repeated change of position, and drawing the sheets straight, are requisite, for *in no disease does gangrenous decubitus come so easily as in typhoid.*   The patient must not always lie upon his back—he must be laid now on one side, now on the other; even a

* Lye : "Water impregnated with alkaline salt imbibed from the ashes of wood."—WEBSTER.

half, or a whole abdominal position may be judicious. All measures to be taken against decubitus previously given on page 78 must be accurately observed.

In the partial, or in the complete, unconsciousness of many typhoid patients, one may wait in vain for them to ask for drink or for food. For that reason the nurse must see to the feeding of the patient. When he is not sleeping quite soundly, some liquid must be given every half-hour. With his generally high-fever temperature, he requires much refreshing drink. As it is certain that typhoid germs are concealed in ordinary drinking water (the bacteriological investigation is able to prove the existence of bacilli, even in water found quite pure * by chemical analysis), so, where possible, the use of natural mineral waters for drinking is recommended. Those about a typhoid patient, as well as the nurse, should also drink them. To kill the germs, boiling the water is not sufficient. At regular periods, beef-tea, milk, eggs, and wine must be given to the patient. As only a few tablespoonfuls of liquid can be taken at a time, often only by teaspoonfuls, this kind of food must be given many times daily—in severe cases, must be continued for weeks together. The thorough drying-up of the mouth, and the tongue becoming fissured may thus sometimes be avoided.

In severe, prolonged cases of typhoid one nurse is not sufficient—several nurses must relieve each other.

SERIOUS CONTINGENCIES IN TYPHOID. NECESSARY CAREFUL ATTENTION OF THE NURSE.

The nurse must immediately tell the doctor of any remarkable temperatures (*over* 105·8° F. = 41° C., and *under* 96·8° F.

---

* These germs are not to be detected by chemical, but only by microscopical investigation.— R. G.

= 36° C.); how often the temperature is to be taken, the doctor must determine.

The stupefied *(typhous)* condition may increase to *complete unconsciousness* (especially in the third and fourth weeks); *deliria, with great restlessness and frenzy*, may also be present without precisely indicating imminent danger. The nurse must not be left alone with a raving patient who is no longer to be appeased by being quietly spoken to, for she will not be able to keep him in bed, or to prevent him from escaping from the room, or from jumping out of window, or seizing knives, scissors, and such-like articles that may be lying about, with which he may inflict injury upon her or upon himself. Generally he is soon quieted; whether cold irrigations or wet packings should be applied, or soothing medicines given, the doctor must decide.

Sometimes typhoid patients sink *(collapse)* rather quickly; hands, feet, nose and ears become cold, bluish, with cadaverous expression of face, whilst the blood-temperature may yet be moderately high. Such loss of strength *(collapse)* occurs after hæmorrhages (intestinal, gastric, nasal, or pulmonary), vomiting, severe diarrhœa, perforation of the intestine by typhoid ulcers (mostly combined with sudden pain in the abdomen), and after too cold or too long continued baths, or sitting up too quickly, or for too long a period. This is a condition not free from danger, and may occasionally lead to rapid death, from paralysis of the heart. The nurse must promptly give wine and warm drinks (coffee, milk, tea, beef-tea), and wrap the arms and legs in hot flannels, placing hot bottles in the bed; if improvement does not quickly begin, the doctor must be called. Should the dejecta contain blood, they must be kept for his inspection; if the abdomen is painful and no collapse present, cold compresses must be applied to it. It must be an inviolable rule with the nurse to note and make known to the doctor, in as few words as possible, everything unusual, or that deviates from the ordinary course of the disease. A short, accurate report without embellishment, is most to the purpose.

In *hæmorrhage from the nose* the nurse must cause the patient to draw up into the nose vinegar and water (mixed in equal quantities); if it still continues, the doctor must be fetched.

Thrush (Aphthæ) signifies small milk-white spots on the mucous membrane of the mouth and throat, and upon the tongue, caused by a kind of mould-fungus. On seeing these spots the nurse must direct the doctor's attention to them. The mouth must be frequently rinsed with a gargle prescribed by the doctor, the white spots must be rubbed off with a spatula or a dry white sponge, as well as painted frequently with a solution of borax and water (1 to 20).

Impending *inflammation of the lungs*, or of *the larynx* often declares itself by difficulty in breathing, coughing, and expectoration; the last may be tinged with blood, and must be reserved for the doctor. With somnolent patients, already very enfeebled, inflammation of the lungs may come on without these symptoms.

Not only when there is *swelling of the feet* must the urine be kept for the doctor's inspection, but it must be shewn to him frequently—daily when possible. Reports of the quantity of liquid taken, and of the urine discharged in the twenty-four hours, will always be welcomed by him.

Purulent discharges may issue from the *ear ;* these are at first recognised by the somewhat bad odour and the stains of pus on the pillows.

In typhoid, the *parotid gland* is often inflamed on one, or on both sides, and declares itself by painful swelling and pain on opening the mouth. In various parts of the body, swellings (abscesses) passing into suppuration may appear; these do not always produce pain, and are often noticed by the nurse sooner than by the patient.

As the doctor cannot thoroughly examine the patient's whole body at every visit, his attention must be drawn to these appearances. Spots and pimples—in considerable number and intensity—may appear upon the skin, even in non-contagious enteric fever.

TREATMENT OF TYPHOID FEVER WITH BATHS AND COLD WATER.

Sometimes the typhoid poison exercises so rapid a decomposing influence upon the blood that all treatment is in vain; but this is not often the case. Most persons are able to overcome the disturbances caused by the typhoid poison in the blood when the lowering of the accompanying continuous high temperature of the blood and of the body *(the fever)* has been successfully effected; and when the respiratory movements have become vigorous, the functions of the nervous system properly fulfilled, and when, by nourishment, the strength has been maintained until the typhoid poison is expelled from the body, and the organs that were diseased are again duly performing their functions.

Among remedies calculated to reduce fever, cool baths and other ways of applying cold water, have had the greatest success, not only in preserving life, but in preventing serious contingencies and secondary diseases, as already named.

Although the doctor always gives the general instructions for this treatment, yet, in carrying them out, the nurse has very much to do—hers is a heavy responsibility; but *in such cases she has the right, and the satisfaction, of being able to claim for herself a great part of the success obtained.*

A typhoid patient, regularly treated with cold baths by an experienced hand, presents quite a different appearance from one not so treated : from his looks, and the condition of the tongue, lips, &c., the doctor sees whether the treatment has been properly carried out or not. The baths can be given in different ways—the various methods of bathing differ from each other in their effects nearly as much as different medicines do. The mere lowering of the fever-temperature by the cold baths is by no means the most important object.

Cold full baths of 68° F. (= 20° C. = 16° R.) are frequently ordered *for adults*—the same water can be used for the same patient several times in succession; meanwhile the bath remains filled, and the temperature of the water, which is nearly that of the room, can be easily again raised by adding some hot water. Time in the bath, as a rule, should be from five to ten minutes. If weak patients are much affected by the bath, keep cold for a considerable time after it, or collapse, then it is judicious to limit its duration to three or four minutes. Baths so shortened operate much more beneficially than tepid baths of far longer duration. Directly after the bath the patient must rest; he is therefore, without drying, wrapped in a dry sheet in the bed, which bed, specially at the foot, should be somewhat warmed; then, lightly covered, and a glass of wine given him; after a time his shirt should be put on. It is advisable to give some wine to weak patients before the bath. With very weak patients, baths can be begun with a temperature of 95° F. (= 35° C. = 28° R.), and, by adding cold water, they can be lowered to 72° F. (= 22° C. = 18° R.) and less.

Some doctors prefer the full bath to be given as follows :

a bath so large that the patient can sit in it with shoulders under water, and beside it a pail of iced or spring water (2 gallons). The bath is placed parallel with the bed, about a yard distant, with a screen between them. After the bath has been filled as quietly as possible with water of the required temperature, the screen is drawn away. The patient is then lifted into the bath, and immediately a few quarts of iced water are poured over him (so that he feels the temperature of the bath water to be less unpleasant).

The douching of the head is repeated, at the middle and at the end of the bath, in such a way that the water runs very gently and slowly over it, so as to contribute to its cooling as much as possible. In the bath the patient is gently rubbed, water is given him to drink, and if towards the end his patience should fail, he must be encouraged. After the final douching, the bed having been carefully re-arranged meanwhile, he is lifted back into it, with little or no drying, his shirt is put on, his feet wrapped in blankets, and if necessary, warmed with hot bottles. The previously shut windows are opened, the screen re-introduced, and he is allowed to rest. When bath-water is not soiled by the patient, it only requires to be changed every twenty-four hours (Brand's method).

Instead of a full bath a *demi-bath* may be ordered. For this, the bath is filled to the height of about 9 inches, with water at the temperature of from 65° to 81° F.($= 18$ to 27° C. $= 15$ to 22° R.). In this the patient is placed with a cold wet cap on his head, and is immediately douched with the bath water ; he is then rubbed with a firm hand, and is enjoined to rub himself as well. Time in the bath, eight to ten minutes ; when possible, he should leave the bath with the skin reddened, then be put into bed, rubbed dry, and, if necessary, his body immediately enwrapped in a wet compress.

In the intervals between the baths the application of a *wet compress* in the form of a body-compress is often ordered. For this, a damp sheet, folded three or four times, is smoothly spread upon a dry sheet previously folded in the same way. This compress is passed (like a draw-sheet) under the patient so that its top edges come up to the armpits ; the wet sheet is quickly turned from both sides over the body, and then the dry is folded over in the same way (Winternitz' method). Such body-compresses are changed every one or two hours, according as the doctor orders.

For patients who are already unconscious and have ice-cold extremities, vigorous syringing with cold water against the face, rubbing the hands and feet, and applying hot bottles to the latter are recommended.

Sometimes the doctor prescribes *ablutions* only. For these water is used of 50° F (= 10° C. = 8° R.) on a sponge, more or less profusely saturated, as may be prescribed. First, arms and hands, then face and head, are washed ; then the body and legs are not only washed, but they are also vigorously rubbed. Rubbing dry follows.

The wet cold *packing*, or *swathing* (see p. 124), is done as follows :—A stout blanket is laid smooth upon a bed, or a sofa, over which a more or less (as ordered by the doctor) wrung-out, but somewhat smaller, sheet is stretched. In the middle of this sheet the patient lays himself, naked, with arms and legs straight. Taking one side of the sheet, the nurse lays it close under his chin, covers his chest, lays a fold between the arm and chest, then covers the body, thigh, and leg, laying a fold between the thighs. In like manner the other side is dealt with, care being always taken that the sheet fits closely, is smoothly and equally folded around the neck. Then one side of the blanket, for its whole length, is brought tightly round the front part of the body towards the opposite side, then the parts of the blanket, not as yet closely fitting, are drawn firmly together, and the other side of the blanket is wrapped around over the first side. Thus the patient is completely swathed like a baby in arms. If the feet be cold, they need not be packed, but they may be warmed by rubbing or by the application of hot bottles. The head must be covered with a cold wet cap. As the doctor may order, the patient remains in the packing from ten to twenty-five minutes (when the illness is not feverish, even longer). After the packing, a demi-bath or wet friction follows.

Assistance may be also rendered in feverish illnesses by two or three packings, rapidly following each other at intervals of from ten to fifteen minutes, when the patient is packed in two sheets, the one laid over the other, and without blankets, and this is much more easily done than the foregoing.

Cold friction, like the packing, requires much strength of arm. In the friction, two persons can take part—the one rubbing the upper part of the body, and the other the lower. For the method of damp friction see p. 124.

Lastly, *cross-bandages* are much used, and they are applied as follows: take two body-bandages, 2½ yards long and 8 inches wide, and sew narrow tapes to one only; roll up each separately; one, previously dipped in cold water, carry from the lower half of the left side of the thoracic wall over the front surface of the body obliquely upwards to the right shoulder, thence, "turning" the bandage, across the back to the point of commencement; from this transversely over the chest to, and under, the right armpit, then across the back over the left shoulder, and thence again to the right arm-pit. Then lay on the dry bandage in the same way, so that every part of the wet one is covered, and fasten with the tapes.

CARE OF TYPHOID PATIENTS DURING
CONVALESCENCE.

Typhoid cannot be regarded as having *run its course* until the patient has been without fever for a week.   Recovery (*convalescence*) after severe—at times, after moderately severe —attacks is wonderfully slow.   *The convalescent from typhoid* (in fact, every convalescent from severe illness) *requires to be watched over most carefully.*   Too quickly, or too long, sitting-up, much talking, violent noise, dazzling light, emotional excitement — everything exhausts him.   He must not be permitted to get up by himself, or to go alone to the water-closet—severe fainting-fits, even death from paralysis of the heart, partly from the exertion in walking and partly from violent straining, have not seldom occurred.   Brain and heart take a long time before they are again quite healthy and strong.   Slight delirium without fever, principally at night before falling asleep, is not rare ; sudden hallucinations having irresistible tendency to suicide, terrible fears, which the patients endeavour to shun by jumping out of the window, even complete mental aberration (though this is always transient), may exist for weeks or for months during the period of convalescence from typhoid fever.

In the preparation of diet for convalescents the nurse must be very watchful ; she must keep most strictly to medical orders.   Generally, solid food should only be given when the evacuations are solid.   Notwithstanding, there are types in which the stool never becomes fluid at all, as there are also very severe cases without any fever.   Often great morbid hunger is developed in convalescents—they would go on eating everything given to them ; then disturbance of digestion may have very prejudicial results ; renewed hæmorrhage and diarrhœa lower the strength again to a dangerous degree.

In typhoid, many parts of the body, specially the muscles *(the flesh)* and the adipose tissue, waste completely away from disintegration; these must be formed entirely anew in the period of convalescence. It is often observed that young, delicate persons, become far stronger and fatter after happily recovering from typhoid than they were before—indeed, sometimes they become more lively and more cheerful in their nature and character than they were known to be before their illness.

CHOLERA.

Asiatic Cholera is as little contagious from person to person as typhoid; the morbific poison develops only in the stools and the vomit (both termed medically, *dejections*). Every water-closet into which a non-disinfected cholera-stool is evacuated, may become (particularly in badly-kept hotels, restaurants, and railway stations) the source of infection to many persons; these again infect other closets, and so the disease is spread, the poison being developed by a most minute organism, only visible through very powerful magnifying glasses. Naturally, in districts where people are crowded together, its dissemination is very rapid.

If all persons presented an equally favourable soil for this minute organism, then epidemic diseases (some of which occur in summer) would spread much more widely and with greater rapidity. The cholera-spore germinates more easily and more quickly (two to three days) than the typhoid-spore (two to three weeks): in this respect they are in proportion to each other as cress is to grass. With cholera, the greater danger of dissemination lies in this, that the stools of cholera patients are not only much more frequent and thinner (like rice-water), but patients also vomit very much. From the watery nature of these dejections, the bedding and neighbourhood of the bed are more easily bespattered. With the poor the bedding cannot be continually changed, and so the bespattered matter dries into it, and afterwards, as dust, is eddied into the air ; unless previously disinfected, cholera is thus very frequently transmitted to laundresses with the linen.

From this it is seen, not only how the spreading of the disease may be checked, but also why it is, when impossible to take proper measures, whole families and households fall ill of cholera, and are swept away.

In cholera everything described as necessary to be done in typhoid must be attended to with redoubled energy. It is urgently important to put soiled linen at once into a vessel with lye, and to cleanse the pans and seats of the water-closets with solutions of carbolic acid.

Sometimes cholera begins suddenly with very violent symptoms : frequent vomiting and diarrhœa, accompanied by extremely rapid sinking of strength *(collapse)*, great pain, absence of pulse, muscular spasms, and cessation of all evacuation of urine. With cholera, the evacuation of the bowels has a peculiar quality, and is termed "rice-water stool."

In cholera, hot baths are usually given ; at times alternating with cold applications. To lessen the vomiting, swallowing small pieces of ice and drinking cold carbonated water, or, in default of the latter, effervescible powder in ice-water, have been most approved. Strong wine, cognac with cold soda water, and Hoffmann's drops are also given to the patient, after he has been put to bed and warmed; the doctor determines the further method of treatment.

Cholera occurs in very different degrees—at one time, quite slight and transitory, at another, very severe and fatal in a few hours. If a person has overcome a severe attack, the cholera-virus may have caused organic destruction similar to that caused by typhoid-virus, and a condition like typhoid (*cholera-typhoid*) is developed. The patient must then be nursed with the greatest care, completely in accordance with the rules laid down for typhoid patients.

In nursing cholera patients the nurse must never put her hands to her mouth until she has thoroughly washed them with soap and brush. Before going to other persons from attendance on cholera patients every nurse must thoroughly cleanse herself, changing all her clothes, and must disinfect her linen, dresses, and shoes before re-using them. As cholera is very frequently preluded by seemingly slight, so-called premonitory diarrhœa (but the timely treatment of the latter often appears to prevent an outbreak of the disease), so the nurse must pay attention thereto in herself. Where possible, ordinary drinking water must be replaced by natural carbonic acid waters.

DIARRHŒA. DYSENTERY.

Diarrhœa is an inflammatory disease of the lower portion of the intestine, and generally appears in the hot season from causes not always known. It is non-contagious from person to person, and is probably disseminated by the evacuations in the same way as cholera and typhoid. To spread much there must be, in addition, a crowded state of the population, with very unfavourable conditions of the weather and of nourishment, conditions that may occur in prisons, in hospitals, and in armies engaged in war.

For disinfecting the stools the same course must be pursued as already described. In sudden attacks, when the disease declares itself in very violent pains in the body, with constantly-recurring desire to go to stool, each time resulting in only slight evacuations of mucus, blood, pus, sago-like masses, gangrenous shreds, the patient must immediately be put to bed—the room temperature to be 65° F. to 68° F. (= 18° to 20° C. = 15° to 16° R.). Two beds, for change, are very desirable. The drink must be tepid, barley-water, oat-meal-water, or milk of almonds. If there be no vomiting, and the patient is inclined to eat, then milk, strong soups, and yolks of eggs may be given. Violent pains *(colics)* are to be quieted by placing warm compresses upon the body; if constant, painful desire to go to stool be felt, cold sitz-baths of short duration, and starch enemata with a few drops of tincture of opium, are to be given; all else necessary the doctor must prescribe.

DISEASES DIRECTLY INFECTIOUS FROM PERSON TO PERSON.

To this category belong: Typhus fever *(period of incubation,* seven to fourteen days); relapsing typhus (ditto, five to seven days); measles (ditto, ten to fourteen days); German measles, scarlet fever (ditto, one to two days); small-pox (ditto, twelve to fourteen days); whooping-cough (ditto, two to seven days); diphtheria (ditto, two to eight days); mumps (ditto, eight to fourteen days).

On the nursing of patients suffering from any of these diseases except diphtheria there is hardly anything special to be said.

They are more or less severe, acute diseases, accompanied by fever, and are caused by the reception into the system of the virus (most minute fungus-spores) in the form of dust. In these diseases the virus is produced ever anew in the sick-room, so that its power to infect and to disseminate is very far reaching—it is more astonishing that these diseases sometimes cease than that they spread further ; it is possible, that, after a certain time, the propagating power of the fungus-spores becomes extinct. As man is not their true soil for development, so their successive generations, when obliged to live always on the humours of the human body, become perhaps more and more feeble until their capacity for propagation is wholly exhausted.

Diphtheria, which presents much that is peculiar to itself, we shall consider subsequently. It is as well known that measles, German measles, and scarlet fever more often, and especially, attack children, and are very easily transmitted to them, as it is that no age protects against typhus fever, relapsing typhus, and small-pox. It is also a recognised fact that, excepting measles, persons are not easily attacked twice by these diseases, but that a certain measure of protection has been gained by having had them.

To mothers of families and nurses it is of interest to know somewhat of the way in which measles, German measles, scarlet fever, and small-pox begin. Recognition of these diseases at their commencement is not always so simple as the unlearned believe. All are wont to begin with fever more or less violent; in all, at the beginning, redness or red spots, or nodules, are found upon the mucous membrane of the palate and throat, then upon the skin; in the first hours even the experienced doctor sometimes finds it very difficult to diagnose which of these diseases is before him, when he is not justified, from contemporary cases in his practice, in concluding, with some probability, on the specific illness. It is a rule, that *every feverish child should be immediately put to bed and be kept quiet.*

Generally in *measles*, very soon after the introductory fever, spots appear upon the soft palate, then the eyes are reddened, they weep, and are sensitive to light; frequent sneezing, trickling from the nose, coughing; four or five days after the commencement of the fever the eruption appears in the form of small red spots and most minute nodules, or maculæ, first on the face, passing to the body, the arms, and lastly to the legs.

In *German measles,* the lightest and most quickly passing of acute skin diseases, the phenomena are very similar to those in measles, but the small red spots appear almost simultaneously with the fever, which is mostly slight, sometimes is wholly absent. Children generally are so slightly affected by the attack that it is not deemed necessary to keep them in bed.

Although *scarlet fever* in many cases may run its course very lightly, yet, for children as well as for adults, it is one of the diseases most feared because it is often accompanied by diphtheria. In scarlet fever, the introductory fever (which, in many other epidemics, runs its course very lightly) is generally very violent, and sometimes comes on with such intensity that children succumb to it in a few hours. It is usually accompanied with shooting pain in the throat, severe headache, somnolence, or sleeplessness; at times with vomiting and spasms (convulsions). Ordinarily, spotty redness of the skin appears on the first, sometimes only on the second day; primarily on the neck, chest, hands, thighs, then spreads over the whole body, often attended with violent smarting and itching. Of late, the same treatment with baths, packings, douches, and ablutions has been adopted as in typhoid fever. For this treatment, see pp. 206 *et seq.* Often the troublesome itching is much lessened by frequent washing, and repeatedly changing the linen and the bed. Some doctors order embrocations with lard or oil to alleviate the itching, and these, like the washing, must be done on the different parts of the body in succession. If diphtheria of the nose and throat shows itself, then proceed as with other diphtheria cases. (See Diphtheria.)

With these eruptions, as with many feverish illnesses in children, it is most important minutely to examine the throat *(pharyngeal organs).* With refractory children, persuading or shouting at them is of no use, and a nurse must be inexperienced if she thinks in this way to make an examination possible. With children, the best and quickest way, and that least distressing to them, is to set the child upon the nurse's lap; her legs must hold the legs of the child, and her arms must hold the arms of the child securely; a second person must fix immovably the head of the child, clasping it by both sides, and inclining it backwards.

In scarlet fever—as in many other diseases—it is of great importance for the doctor to see the urine frequently; unasked, the nurse must show it to him—this is best done in glass vessels.

Man is subject to different varieties of *pock or small-pox;* a slight form, *chicken-pox (varicella);* a more severe *(varioloid);* the most severe, *the true small-pox (variola).*

Chicken-pox is a slight illness—fever and other precursors are often absent; small red spots, isolated, are sparsely spread over the chest, neck, face, arms, and legs, and on the second day these become small vesicles, which, on the fourth or fifth day, are again dried up; sometimes there is no fever;

in other cases the temperature is increased, with sore throat, which soon subsides, and with vesicles on the palate and tongue. Children suffering from it are confined to the bedroom ; when not feverish, lying in bed is unnecessary. Isolation is as needful as in true small-pox, for many doctors believe that true small-pox can be acquired by transmission from this lighter form of it.

When *true variolous poison* has been absorbed into the system and takes effect, then, as a rule, twelve to thirteen days elapse before the outbreak of the introductory fever; yet, during this period of incubation, vertigo, languor, pain in the head and loins may be present. Often the disease begins with violent rigor, or repeated chills, with high temperature of the body —with loss of appetite, inclination to vomit, retching, even actual vomiting, increasingly violent pains in the head and loins, delirium also at times, and dragging, acute pains in the arms. In some cases, even at this period of the disease, red measle-like, or scarlet-fever-like spots appear upon the skin, with also, at times, small scattered spots of blood *(petechiæ)*. On the third or fourth day after the fever begins, the true small-pox eruption first distinctly shows itself with abatement of the fever ; the red papulæ, first seen on the palate and the face, then on the trunk and other parts of the body, gradually change to vesicles, and then to purulent pustules.

There is endless variety in the way in which small-pox runs its course—in some cases the number of the pustules is very insignificant, in others so great that the boundary of one pock can scarcely be distinguished from that of another ; each separate pock can penetrate the skin more or less deeply ; it depends upon this whether deep, or superficial, or no scars at all are made. The treatment of small-pox with baths and wet packings has now many medical adherents.

There is one important, but not always absolutely certain, preservative against small-pox—vaccination with cow-pox.* He only who is ignorant of the history of the deadly small-pox epidemics before vaccination became general, and is unable to estimate from it the present greatly lessened danger of this disease, can speak in blind zeal against vaccination. Its protective efficacy is estimated to last for seven years. Nurses and doctors, therefore, should be re-vaccinated frequently; but persons who have much to do with small-pox patients do not easily fall ill.

Whooping cough (choking cough, spasmodic cough, blue cough) largely prevails among children, but it may be communicated to adults. Most probably the infectious matter is contained in the expectorated and vomited mucus, and therefore no one should be permitted to use pocket-handkerchiefs that have been used by such patients. That the infectious matter is in the excretions alone has not been proved. At its commencement, children cough only as in ordinary catarrh, the cough being without any special character. This preliminary stage can last from one to three weeks; then the paroxysms of true spasmodic coughing begin,—these are combined with gasping inspiration, and usually finish with retching and vomiting, when a viscous mucus is expelled. During the paroxysm the anxiety and difficulty in breathing are great, the child often becomes quite blue, the eyes bloodshot, and blood also may be mixed with the vomited mucus.

* Vaccination, or the introduction of cow-pox virus into the human system to make it insusceptible to small-pox (more correctly, less susceptible), was practised by the ancient Indians. In our times and for Europe, the discoverer and scientific founder of the method was Dr Jenner (1749-1823). He was the first to observe that those who were infected by cow-pox were not afterwards infected by small-pox patients.

Rapid eating and "gulping down," screaming, or much running bring on the paroxysms; these, however, may come on frequently in the night, probably from the saliva and mucus of the mouth running into the larynx. Twenty to thirty paroxysms daily are usual; at the height of the disease, from sixty to eighty often occur. As a rule, the whooping-cough-stage lasts from four to five weeks, followed by its slow abatement during from one to two weeks.

Suitable medicine is a question for the doctor. During the frightful coughing one would like to help the child, but then very little can be done. The head must be so held as to facilitate retching and vomiting, and the mucus must be taken from the mouth with a handkerchief. Older children must be ordered to refrain from screaming, raging, and everything that can provoke a paroxysm. Lying high in bed at night is desirable. From observation, adults learn how to shorten the paroxysms by taking, at their commencement, the least possible, and very superficial, inspirations. In the attack, if suffocation be imminent, quickly put the finger far into the patient's mouth, and press down the root of the tongue. Generally the disease runs its course without fever; hence it is better not to keep children in bed, but, in fine warm weather, in the daytime to let them be much in the open air without permitting them to get heated by running, etc.

In nursing *diphtheria*, a special, voluntary sacrifice of self is necessary. This disease, appearing sometimes alone, at others in conjunction with scarlet fever, and more rarely with measles and typhoid, has spread in the last seventy years in Vienna, but nothing like to the same extent as in some other great cities (Berlin, Paris, St Petersburg, London). No period of life is secure from its attack; neither the cottage nor the palace is protected against it, but children of families crowded together in bad, damp dwellings, in which two or three often sleep on rotten straw in one bed, are most frequently attacked. Under such deplorable circumstances the infection spreads most rapidly, and the badly-nourished, neglected, dirty children resist it the least.

Here we shall only treat of diphtheria in the throat, larynx, and nose, wholly excluding that arising from wounds. In this disease also the infectious matter consists of most minute fungus-seeds (spores), visible only when most powerfully magnified. With great facility these find the most suitable soil for germination in the throat, on the soft palate, and on the tonsils. In the beginning it is confined to the points first infected, and often these cannot be recognised as at all morbidly affected; but in a few hours, or it may be after some days, fever and other symptoms appear, showing that the poison, evolved by the fungi in their growth, has penetrated the blood; there are cases in which no such symptoms occur, and the disease remains purely local. As already observed, the infectious matter affects children most easily, even when the mucous membrane is healthy; and adults also, more especially when the mucous membrane of the throat is in an unhealthy condition.

The secretions of the mucous membranes attacked by diphtheria are infectious in a high degree; by spattering in coughing, by direct contact in kissing children ill with it, the morbid matter is conveyed to other members of the family. Even when dry, the mucus dejected into the handkerchiefs, or spattered upon the bedding, *sometimes* retains its infectious power, and, imparted to the air and inhaled in large quantities, can *probably* reproduce the disease. In this way the disease would never come to an end, and its dissemination would be far greater than it is if, in numberless cases, the infectious matter were not rinsed out of the mouth, or swallowed, digested, and evacuated before it had infected ; or if, as with all infectious diseases, other special personal conditions were not necessary to favour the infection by the contagious matter, and its further development in the blood. In what this personal susceptibility, or personal resistance to infectious matters consists, is not known. Many doctors and nurses (mothers, relatives, and professional nurses), in the discharge of their duties, have fallen victims to infection from diphtheria ; but it cannot also be denied, that many imprudences were committed, and that these took place more frequently when the danger of direct contagion from mouth to mouth was not sufficiently known. Were the power of infection in the dried diphtheria matter, when scattered like dust, particularly great, then the dissemination would be far greater, and would extend more particularly to the doctors and nursing staff ; this is the case in a far higher degree in typhus fever and in relapsing typhus than in diphtheria.

The general precautions necessary for guarding against infectious diseases will be subsequently considered. With diphtheria, however, we lay stress upon it here that, from the positive knowledge already gained of the infective power of the infectious matter *(the contagion)*, much can be done to protect those who must come into contact with such patients. Above everything, *direct contact of the mouth with the mouth of the patient must be strictly avoided*, and the precaution must be taken not to let any of the expectoration, or of that which is scattered like dust by the patient in sneezing, get into one's own mouth ; should this happen, then the mouth must be immediately and frequently rinsed out with a solution of salicylic acid (1 part to 300 parts water), or permanganate of potash (from 2 to 3 granules dissolved in a glass of water), or chlorate of potash (5 to 100). Even when not exposed to such direct contact, it is advisable, if compelled to be with diphtheria patients, to rinse the mouth every two hours.

It is of great interest to lady-readers to know the symptoms by which the disease, at its commencement, may be recognised; for, in a disease which first attacks the body in situations so visible as in diphtheria, it is evident that, to oppose it in its first beginning is possible, and of great importance.

Distinctly pronounced cases generally begin with white spots upon the palate, in the throat, or on the tonsils; moderate, shooting pains in the throat sometimes precede these symptoms by a few hours; premonitory fever may be wholly wanting—even in severe cases it can be slight; the spots, at first silvery white and nearly transparent, soon become cloudy-white, milky, and spread, the surrounding mucous membrane becomes darker red, and the white patches peel off in more or less thick membranes. Whilst these cases are proportionately easy to diagnose, there are some which, to recognise as diphtheria, is very difficult, it may be, even impossible, because, with the penetration of a small quantity of infectious matter (feebly infective), into the throat of a person but little predisposed to it, the whole disease may be confined to a moderate, spotty redness of the mucous membrane of the throat, with slight shooting pains, and it is possible that, even later, no white spots appear at all; after a few days of seemingly ordinary catarrh-like inflammation of the throat, the patient is again quite well, or else, the secondary conditions of blood-poisoning arise. When persons who have been in intercourse with diphtheria patients have these symptoms, there is always great probability that they have been slightly infected.

The decision, whether such inflammation in the throat
originates from diphtheria, or from a cold, is somewhat a
matter of indifference to the person attacked, but is not to
those surrounding him, for such slight cases are no less infectious than the severe. Prudence therefore demands that
patients suffering from pharyngeal catarrh and catarrhal diphtheria must be restrained from contact with others, especially
from mouth to mouth contact, just as much as those suffering
from declared spotted, membranous diphtheria; they should
use gargles plentifully, not only to render their slight ailment
harmless to themselves, but to prevent the infectious matter
from being transmitted to others. All the more precaution is
necessary in these cases, because, even after several days, the
worse forms of diphtheria may yet be developed from an attack
of continued, unopposed catarrhal diphtheria, which, perhaps,
was combined with other injurious influences, such as much
speaking, screaming, or a chill; and, because further, the
simple redness of the throat, without visible spots, does not
preclude the dreaded white spots and membranes from being
at the top of the throat behind the soft palate, and in the
back of the nose, and from passing thence, but later, over
to the front side of the soft palate. In scarlet fever, this
concealed seat of diphtheria occurs very frequently.

To cleanse the throat, and *carefully to remove the membranes
which proceed from the white spots, and are inclined to rapid
putrescence, is the principal duty of nurses.* The means by
which to promote this consist in cleansing and painting the
parts diseased, and in making the patient rinse his mouth
very frequently without straining the organs of the throat by
immoderate gargling movements; by injections into the nose,
to wash away its secretions; and by inhalations of steam, to
soften the tissues which are liable to become rigid through the

P

disease, and by which the detachment of the membranes is promoted. The use of these remedies is considered in Chapter III.

In diphtheria it is very necessary to guard against injuries and hæmorrhages being caused by these operations; all instruments used must be kept in three per cent. solution of carbolic; after use, must be re-laid in it; the pieces of sponge and cotton-wool used must be immediately burnt.

The medicaments for the gargles, paintings, and injections must be ordered by the doctor; — here only a few further observations will be made. Remedies are not wanting by which the disease may be quickly overcome at its commencement—the most powerful are very corrosive; in dabbing or painting, they must be used with the greatest care, on a well-squeezed camel's-hair pencil, or a small piece of sponge, for if any of the corrosive matter should run into the larynx, or into the trachea (which may happen more easily with organs when rigid and very checked in their movement than when in health), then a very dangerous inflammation of these parts would ensue, and a new peril would arise. For gargles, and for injections into the nose, as for inhalations, only weak solutions of such remedies must be used as will cause no injury should some get into the trachea (this is always the case in inhalations), indeed, in many cases, some should enter. With little children, local treatment presents considerable difficulty,—at times insurmountable. If, with adults, some skill is requisite to cleanse the space behind and over the soft palate (where it enters the nasal cavity from behind) with a small piece of sponge, or with a pledget of cotton-wool held by a curved forceps, and to instruct an adult that he must so depress the tongue backwards during injections into the nose that the fluid does not run into the throat and windpipe, but simply out of his mouth—

with children, this is scarcely to be attained at all. As only weak solutions of effective remedies can be used, so success cannot be gained but by their *very frequent application—i.e.*, at intervals, from half an hour to one or two hours, according to the severity of the case. Many doctors are satisfied with frequent gargling, and do not require the parts affected to be painted or cauterized, therefore, exact instructions must be given to the nurse. Much gargling is very painful to the patients ; in a great number of cases, it is the mother who gives this remedy, many soon tire, particularly when they see no rapid result ; they also do not like so frequently to give pain to their darlings ; they lose all hope, and soon sink into inactive grief and lamentation. Until the last breath of the patient it is the duty of a capable nurse, as it is that of the doctor, to make every effort experience teaches *may* render help, although in this, as in many other processes of disease, their most devoted endeavours find their only reward in the consciousness of duty faithfully fulfilled.

Should a case of throat and nose diphtheria terminate fatally, then it was the blood-poisoning that caused death (sometimes from heart-paralysis rapidly coming on with symptoms of collapse), unless other diseases supervened, such as inflammation of the brain or of the lungs. But there is still another cause of death which may be sometimes successfully resisted by timely intervention : suffocation from *the transition of the pharyngeal-diphtheria to the larynx and the trachea*, by which these respiratory passages become quite obstructed by the formation of the white membranes already described.

In this case, of which only the doctor can form a judgment, suffocation may be prevented by opening the air-passages *(tracheotomy)*, removing the membranes, and inserting a tube *(canula)*, and by this operation the lives of children are sometimes saved if they succeed in struggling through the diphtheria poisoning, or do not succumb from the diphtheria spreading to the lungs. After the operation, the following duties devolve upon the nurse. The inserted tube, through which the patient breathes until the passage is again free, is easily obstructed, partly by imperfectly expectorated membranes, partly by mucus, which quickly dries in the tube. That this canula may be easily cleansed it is made in two tubes, the one smaller and fitting inside the other; the inner tube is withdrawn, cleansed with a quill-feather, syringed with a three per cent. solution of carbolic acid, and re-inserted. The formation of crust in the tube is prevented by frequent inhalation of warm steam; and the entrance of dust into the tube by fine gauze stretched before it. If, in spite of the frequent cleansing of the canula, the breathing is not free, the doctor must be fetched.

There is a moderately severe form of diphtheria called *membranous quinsy*, or croup, which begins in the larynx, and is accompanied by a peculiar barking cough. By degrees difficulty in breathing, and fear of suffocation, follow; but the throat may be, and may continue, free from diphtheria. Many children succumb to this disease, principally because the inflammation spreads to the lungs. In this disease the danger of blood-poisoning is less than in severe cases of pharyngeal-diphtheria, and the patients are more frequently saved by tracheotomy.

PRECAUTIONARY MEASURES IN EPIDEMICS, AND IN INFECTIOUS DISEASES.

Since more exact knowledge has been gained of the nature of infection, we are able more definitely to determine the necessary steps to guard against its attacks. To prevent fresh outbreaks of epidemics and their transmission, lies, to some extent, in the power of national governments—the absolute isolation of all foci of disease, in this age of railways and personal liberty, is not possible. If, in spite of this, it is proved that epidemics of the worst pestilences in our times have never caused such ravages as the plague and the black death of the Middle Ages, this may be attributed chiefly to the more healthy planning of towns, as well as to the more rational measures which governments, as well as individuals, take to destroy those minute plants, poisonous to us, which produce these diseases. In all nature, being against being struggles for existence and for food. On the whole, in the long run hitherto man has always been victorious, and, for that reason, he is termed the lord of creation.

Assuming that a man is able to choose, and is desirous to continue in health, he will not select a dwelling in a malarious district. Those persons whose means permit will leave a town from the soil and ground of which typhoid poison is abundantly developed, or, once introduced, thrives well, and frequently comes to the surface. He who would escape the possibility of an attack of cholera must absent himself from the town so long as cases of cholera occur, or until the pestilence is extinct ; he who would avoid infection from measles, German measles, scarlet fever, small-pox, diphtheria, whooping-cough, must shun houses, and especially apartments, in which persons are suffering from such diseases. All this is as self-evident as is the advice to a man to protect himself from being shot by not going where the balls are whizzing.

It is very different with those whose circumstances compel them to have intercourse with such patients : the doctors and nurses. What is most important upon this point has already been said in treating of the various individual forms of disease,—but here a short summary shall be given.

Infection always results most easily from the still damp secretions, expectorated matter, and evacuations of the patients. Persons in their immediate vicinity must beware lest any of such matters come in contact with their mucous membranes (mouth, nose, eyes), or with slightly-wounded places on their skin. This specially applies to expectorated matter from typhoid, cholera, diphtheria, and whooping-cough patients.

I will now mention that for which no previous opportunity presented itself, viz., the danger of transmitting pus, sanies, lochia, and diphtheritic secretion to *wounded places in the skin, or to the pores*. With incredible rapidity these matters penetrate as far as the deep cutaneous tissues. During operations and dressings many surgeons have been infected in this way by septic and disease poisons, and have been sacrificed to their vocation. If these poisons enter the blood a very malignant, purulent fever is produced (sepsis, pyæmia), a general blood-poisoning, which is seldom overcome. When, as doctor, one cannot avoid operating in ichorous or suppurating cavities, in every case it is prudent, not only to besmear the hands with oil, but also to rub the oil into the skin. If there be wounds on the fingers, or small cracks round the nails (particularly favourable to the entrance of poisonous matter), cover them with court-plaster, over which spread a thick layer of collodion before doing anything with the hands to surgical patients. Then rub the hands with oil.

If, during an operation, a nurse is accidentally injured, the wound must continue bleeding in water from a quarter to half-an-hour—then wash it with three per cent. solution of carbolic, and not till then cover with court-plaster. Cooks injured in handling meat no longer fresh (chiefly game) must proceed in the same way. If, in spite of this treatment, the flesh surrounding the wound should swell, become sensitive, and the pain extend up the arm, a doctor must be consulted immediately.

Cancer is non-contagious ; yet, under certain circumstances, one may be infected by the secretion of putrid cancerous ulcers equally as by other putrid fluids.

Pulmonary consumption also appears capable of transmission from person to person, although such transmission is more difficult, and comparatively rare. It is therefore not advisable for young people to sleep in a room with a consumptive person, because, from recent careful research, it is deemed highly probable that the sputum of such patients contains matter which is capable, in certain circumstances, of reproducing in others the same disease.

If we consider the sources of infection in epidemics, *it is proved beyond doubt that in typhoid, cholera, and diarrhœa the infection proceeds from water-closets into which the infectious evacuations have entered.* As, generally, by sewers, all water-closets in towns are in communication, and it is impossible for the sewers always to be so supervised that, at no time, could infectious matters pass from them into the soil and thence into the wells, it is evident how difficult, indeed, how impossible it is, under such circumstances, to cut off this source of epidemic disease. Doubtless the most effectual way is, to destroy the infectious matter in the evacuations before pouring them into the water-closet. When this is done, in all hospitals and private dwellings, much will have been accomplished to prevent the spread of epidemics.

There will still remain cases in which persons, unaware of the illness with which they are attacked, and before taking to their beds, pass their evacuations into water-closets without their being disinfected ; but, as it is the *quantity* and *increase* of infectious matter diffused in the sewers that is principally of consequence in disseminating epidemics, so it is of great importance that this source of injury should be diminished by direct disinfection of the evacuations of those under medical treatment. It is further important, that the body of water flushing the sewers shall be so large, and have so great a fall, that the evacuations shall be quickly carried away from the town into a near, rapidly flowing river ; on the one hand the infectious matter becomes so *attenuated*, and on the other hand, so *changed, by the water*, as to be no longer active.

Attempts to destroy infectious matters in water-closets and in sewers have also been made by pouring chemical substances into them (carbolic acid, sulphate of iron, chloride of lime), but the value of so doing is disputed, for it is said the quantity of disinfecting substances used would never be in sufficient proportion to the bulk and unequal composition of the sewer contents so as to annihilate the vehicles of disease, the fungus spores, which are always difficult to destroy. This is generally correct ; but even to retard the germination of these fungi until they are carried further away by the water, will suffice ; and this, in fact, is possible by disinfection carried out daily by sanitary officials during an epidemic, assuming that all water-closets are regularly flushed, and the fall into the sewers is sufficient.

Wells can become centres of disease only when they are badly lined, are not very deep, and contain little water. The fungus spores which carry the disease, remaining in pure water, lose their infective power even after a few days ; they can, however, maintain themselves in the slime adhering to the walls and floor of a well. Therefore, the less water there is in a well the more easily the mud is stirred when the water is drawn, and all the more quickly the well-water may become a source of disease.

Stress must once more be laid upon the fact, that it is not the malodorous gases, as such, that are vehicles of disease ; on the contrary, the more some such gases are evolved (ammonia, sulphuretted hydrogen) all the more the morbific fungi perish ; yet, by exhalation and evaporation (principally by simultaneous formation of bubbles and their bursting), many fungi, still capable of germinating, float into the air, and, from their infinite lightness, are long suspended ere they drop and become dry ; whilst in the air, man can inhale them with the malodorous gases, and thus they get into the blood.

Infection from perfectly dust-dry morbid matter takes place much less easily—indeed, in cholera, typhoid, and diarrhœa it is doubtful if it occur at all ; but in measles, German measles, scarlet fever, small-pox, and diphtheria, this source of infection does exist beyond all doubt.

In the acute skin-diseases mentioned it is as yet not very clear where, in the body, the infectious matter is mostly concentrated. Of small-pox, however, it is positively known that the morbific matter is either absorbed in the form of dust by respiration, or is propagated (inoculated) by transmission of the contents of the small-pox vesicles into the skin of another person ; even by the latter method of intentional infection (not carried out with human, but with cow-pox, see p. 220), not only does the part of the body, artificially made diseased, become affected, but the matter passes into the blood through the whole body, as is proved by the not altogether trifling fever which sometimes sets in in those who have been vaccinated. We also know that the small-pox virus, even when dried in the vesicles and scabs, is still present in the scabs, and these do not disappear without leaving their traces in the form of dust. We know further, that operative inoculation *can* be made with a mixture formed by stirring the said scabs in water; but *such inoculations are rarely effective*, although the most favourable conditions for taking are produced by conveying the matter *into* the skin by a lancet —which, with the natural method of small-pox infection, never happens.

As, from this it is evident, at any rate in small-pox, that the infectious matter is in the skin and comes to its surface, so, from its great analogy to *measles, German measles, scarlet fever, and diphtheria* in their other modes of appearing and spreading, it is in the highest degree probable that, *in these latter diseases* also, the infectious matter is seated in the skin, thence comes to the surface, perhaps exudes with the perspiration, at all events, finally passes in the form of dust from the surface of the skin to the linen and clothes, and into the air surrounding the sick.* As these skin diseases always terminate with a thorough, dry scaling of the epidermis, so, and with reason, such patients are always deemed infectious until the desquamation is finished, some baths taken, and the linen and clothes changed.

In these diseases, although the infectious matter is certainly in the skin, it does not follow that it cannot also be found in other parts of the body; it is therefore necessary to guard against contact with all evacuations and secretions from such patients, and to disinfect them as described.

When the dust-like infectious matter loses its germinating

---

* To these Erysipelas may be added, although it possesses only a very slight and very conditional power of infection.

power, whether, in fact, under ordinary circumstances upon the earth's surface, it ever loses it at all, is not known.   As it is a most minute, dry germ, so it may preserve its life, though dormant, just as long as do the grains of wheat which the Egyptians deposited in the Pyramids thousands of years since, and which, sown in the soil to-day, germinate and bear fruit as if only taken from the ears of corn grown the year before. Degrees of heat and cold, which would destroy most substances of which animals and plants consist, have no power to annihilate the vitality, the power to germinate inherent in these finest fungus-spores.

Now where, and how, can one cope with this injurious dust?

When we are ill this injurious dust can be in the surrounding air; on us personally, on our skin, in the hair; on the furniture; on the walls of the room ; in our clothes ; in our linen—how shall others escape from it? how guard against it?

Before describing the practical measures to be taken to prevent infection from this source, and to allay anxiety, it is very important to note as follows :—

I. In order that dust-dry fungus-spores shall reach development they must adhere to some part of the body for a time, where aqueous matter for swelling and germinating amply exists.   On the dry surface of the body these dry spores cannot germinate at all—with the air *only* can they gain admittance *into the mouth and nose*, and thence into the lungs and stomach.   The ways of access, therefore, are very small in proportion to the surface of the body.

II. It is very probable that, upon the whole, most of the disease-carrying germs do not reach germination at all in the acid gastric juice, in the gastric mucus, or in the chyme ; possibly they settle rather in the lower part of the intestine, particularly at small sore places.   We know that the species of small fungi which produce decay, and grow exuberantly to thousands of millions in an atom of piquant cheese, do not hurt the stomach or intestine. In spite of this, whether the fungi (unusually tenacious of life) of typhoid, of cholera, and of diarrhœa, which get into the chyme and the intestine, there develop, and thence enter into the blood with the absorbed chyle, is not known, but it is possible; in all cases it is advisable, in epidemics of cholera, typhoid, or diarrhœa, not to change the mode of living, but to eat only newly-prepared, well-cooked foods, to avoid indigestible foods, and raw, un-cleaned fruit. It is advisable also, *never to remain long without food*, so that the stomach and intestine do not become perfectly empty ; to take good red and port wine is to be recommended.

Of the remaining germs of disease here in question it is at least highly probable that, only by respiration do they enter the blood through the lungs. The entrance doorway is there-fore very limited, although mostly open : the air flows in with every inspiration. The germ of diphtheria cleaves very easily to the mouth, specially to the unclean mouth, at places little affected by movements in speaking, masticating, and swallowing. Wounded persons, if the infectious matter reach them, are easily attacked by scarlet fever.

III. Therefore of great, practical significance is the fol-lowing :—From the continuous compensation of tempera-tures the air-movement on the surface of the earth is very considerable, even when no wind is noticeable ; and the fungus spores, being inconceivably light, are very rapidly scattered in space in every direction, and this distribution soon becomes endless. Researches have been made from which it appears that a quart of fœtid air of a mortuary contains about ten spores of septic fungi ; whilst in one

drop of putrescent liquid, so many fungus-spores luxuriantly
swarm that the term "milliards" conveys no adequate con-
ception of their number. If therewith we take into consideration how easily the conditions are presented for the further
growth of fungi, already vigorously and exuberantly growing,
and, on the contrary, with what difficulty all the conditions
are combined which bring the dust-dry fungus to germination, it follows that damp infectious matters are much more
perilous than dust-dry. This has been known for a long
time—no one fears to pass a house in which a scarlet fever
patient lies; anxiety begins on setting foot within the doorway,
and increases step by step to the sick-room, and until in close
proximity to the patient; in this, popular sentiment is correct.

The more frequently the air of the sick-room is changed
by vigorous ventilation the better, not only for the patient, but
the more the infectious matter is attenuated so much the less
is the probability of infection for the persons around him;
so much less the accumulation of infectious matter upon the
furniture, in the bedding, in the clothes; and so much less the
probability of the rest of the inmates of the house being
infected. I refer to what has been said on this subject in
Chapter I.: to let no dust rise; to remove it with a damp
cloth so as to prevent it from again eddying into the air; and
occasionally to lay it by artificial rain (spray).

Yet a few words upon *disinfection.* Disinfection signifies
the freeing of the air, clothes, linen, beds, surface of the body,
&c., from infectious matters, their annihilation being the
object aimed at.

All former methods of disinfecting clothes, linen, sick-room,
&c., accomplished little—indeed, not knowing the nature of
the infecting matter, persons were not in the position to be
able to say whether the process of disinfection adopted was

really successful or otherwise. When, after such disinfection, —for instance, of a room in which a scarlet fever patient had lain,—no further case occurred in the house, it might be said that such cessation of the disease very often happened without any disinfecting. As already observed, it is only since we have known the micro-organisms to be the carriers and generators of many, specially of infectious, diseases, that it has been possible to determine which process of disinfection has power to annihilate them, and those forms of them most capable of resistance, the "spores," and with them, as may be foreseen, the diseases dependent upon them. Experiments with these micro-organisms have been made on a large scale in the German Imperial Health Bureau, in Berlin, and their results have become a standard for modern methods of disinfection. The course pursued was as follows: morbific bacilli were cultivated, such as the well-known anthrax bacilli (which produce spores very tenacious of life), easily recognised under the microscope; threads of silk were saturated with them, and, after subjection to various processes of disinfection, were tested to see if the anthrax bacilli could be further cultivated or not—*i.e.*, whether they were killed or not. It was proved that anthrax *spores* were killed only after being subjected for seven days to a three per cent. carbolic solution, but the anthrax *bacilli* were killed after two minutes' influence of a one per cent. solution of carbolic acid. Perchloride of mercury kills the anthrax spores even in a solution of one to five thousand, but this may not be used, because of its poisonous character. The perchloride and the carbolic acid are often ineffective, because they enter easily into insoluble combinations. All micro-organisms are not killed by hot air, even when its temperature is raised to over 212° F. (= 100° C.) Were the air heated to

284° F. (= 140° C.) for three hours, all germs would be killed ; but it was found that the temperature in the inside of blankets exposed to this heat reached only 158° to 199·4° F. (= 70 to 93° C.) and, consequently, the fungi therein were not all destroyed—only when a current of *steam* at 212° F. (= 100° C.) was used for at least an hour was the result satisfactory.

As a consequence of these experiments, disinfecting ovens, with steam circulating through them, are now used, such as is in the Rudolfiner Haus, Vienna. The articles remain in the oven from one to one and a half hours, and are taken out uninjured. Many parishes are now providing such ovens.

Disinfecting with chlorine or bromine vapours entails enormous difficulty, and causes much damage in the places thus disinfected ; the kind of articles to be disinfected must always be taken into account.

Where arrangements for such disinfecting processes are not available, the articles must be dealt with in accordance with directions already given. The safest course is, to burn everything consumable, but the value of the articles often prevents this. Unpolished metal articles, paperhangings (if not renewed), the walls, &c., are best treated with a five per cent. solution of carbolic acid ; floors and wooden articles washed with a solution of perchloride of mercury (1 : 1000), followed by rinsing.

Furniture, carpets, cushions, clothes, must be beaten in the open air, and the person who beats them should tie cotton-wool before his mouth and nose so that the air inhaled is filtered. If the sick-room had been always well ventilated, and the articles that came into direct contact with the patient were often changed and cleansed, then this will suffice for their disinfection. Fumigations with chlorine vapours, or with sulphurous acid, and the use of dry-heat in specially constructed disinfecting ovens (now practised in many hospitals), kill the fleas, &c., but do not destroy the germinating power of infectious matters; if the heat and thorough fumigation processes with the said vapours were carried so far that the latter object was attained, then also nothing would remain of the clothes, &c.; in that case, rather let them all be burnt. For the rags worn by the poorest hospital patients all such processes are unnecessary; if their clothes do not bear washing, destroy them, and give them new on leaving. Apart from the benefit this would be to the individual, it is the best means for protecting the public, who otherwise might be infected subsequently by such clothes.

CHAPTER VII.

CARE OF NERVOUS PATIENTS, AND PATIENTS MENTALLY DISEASED.

BRAIN, spinal cord, and nerves are all in uninterrupted connection, and they form the nerve-apparatus *(nervous system)* of the body.   By it we not only become acquainted with the external world—because, with its assistance we see, hear, taste, smell, and feel,—but it also causes motion, partly *without our will* (movements of the heart and intestines : *involuntary movements)*, and partly *by our will (voluntary movement).* The whole apparatus must be vigorously constituted, in complete undisturbed connection, and it must be well sustained in its functions if it is to continue to work, regularly, without interruption, throughout a long human life ; for this, it is essential that it should be perfused with a sufficient quantity of healthy blood. That this blood shall continue healthy, the nourishment must be suitable, and the action of the lungs sufficient. To secure this it is necessary that the digestive and respiratory organs shall be healthy. Finally, the heart, by its regular (rhythmic) drawing-in and expelling of the blood, must impel it through the blood-ducts *(blood-vessels)*, so that all organs may continuously receive blood refreshed by the respiration. The heart is also affected by the power proceeding from the brain, which flows to it through specific nerves. The task of the brain is, not only to receive as sensations what is going on outside of it, and to send out motive power, but it is also the seat of the ideas, of the "soul."

These explanations may be difficult to comprehend, even by the educated laity. Here they shall only indicate how extremely complicated is the process briefly termed "life," and from how many causes the activity *(function)* of the nervous system can be deranged.

The derangements are chiefly of three kinds—derangements of the *sensory*, of the *motor*, and of the *mental, functions* of the brain. The function of the nervous system is—

Fir st—*Morbidly increased*, and shows itself

    1. In the region of sensation—as increased sensibility, pain.

    2.   ,,   ,,   motion—as cramp.

    3.   :,   ,,   the mental function—as hallucination, delirium.

Second—*Enfeebled*, or *wholly suspended*, and shows itself
    1. In the region of sensation—as insensibility.
    2.   ,,   ,,   motion — as partial or complete
                paralysis.
    3.   ,,   ,,   the mental function—as imbecility,
                idiocy.

At all events, under all these conditions, the nervous system
is suffering, yet it is not necessary that the first cause of this
suffering shall lie in the nerve-apparatus itself, but the constitution of the blood and the way in which it is distributed
may also be responsible. If the derangement originating in
the blood and the vessels be transitory, then the derangement
of the function of the nervous system will be also transitory ;
if it increase, or be permanent, then the disease of the nerves
will also increase.

Intimately as the diseases of the nerves and the mind are
combined, for practical reasons, it is still useful to separate
what there is to say on the care of patients affected by them.

NURSING AND HELPING IN NERVOUS DISEASES, AND IN ATTACKS ORIGINATING CHIEFLY IN THE NERVOUS SYSTEM.

Acute inflammations of the brain and spinal cord, combined with violent fever, and running their course rapidly, are
rare in adults; of course, after injuries, these diseases may
arise at any age. On the nursing of such patients nothing
special need be said—the same principles apply as with all
feverish patients. In the diseases of the said organs which
run their course slowly (chronic), constant nursing is, generally,
only necessary when paralysis comes on. This may arise
gradually or suddenly. If paralysis of the legs is complete, so

that the patient must continually lie down, and, if it is
followed by paralysis of the sphincter muscles of the bladder
and anus, so that these unhappy patients soil themselves unconsciously, then careful sick-nursing is able to do much to
preserve life, except in cases where rapidly progressive disease
of the brain is the cause of paralysis. If the brain is un-
affected, and the spinal cord alone suffers (from injury or
disease), the chief danger then threatening the patient is from
lying bed-sore, mainly from gangrenous decubitus. In such
a case, all instructions previously given upon this subject
(p. 78) must be most minutely observed. Paralysis of the
sphincter muscles is seldom so immediately complete that a
certain quantity of urine and fæces is not retained. If, however, the urine be evacuated by the catheter about every three
hours, the patient may sometimes be kept tolerably dry. Only
when in a thin, almost liquid state, are the fæces involuntarily
evacuated by these patients. If hard (more frequently the
case), they must be removed, not only by irrigations, but also
mechanically, in accordance with the doctor's orders.

As a precaution *(prophylaxis)* for the prevention of decubitus
it is important, not only to remove these patients frequently
from one bed to another, but, where possible, to keep them
for a while in a sitting-posture in an arm-chair, so as not always
to expose the same places to pressure from the weight of
the body. From these changes only those are excepted who
are injured in the spinal cord and in the vertebral column—
they should not be moved much, and should be laid upon
water-beds (p. 81) as soon as possible.

A series of attacks occurs, proceeding partly from more or
less rapid and transitory, and partly from enduring, changes
in the function and constitution of the brain—in such cases,
instant help is not only desirable, but is urgently necessary.

Of these attacks *fainting* most frequently occurs. With the sensations of giddiness, nausea, and weakness, unconsciousness comes on; the fainting person falls down, the face is deadly pale, pulse very small, respiration superficial and slow, the limbs and whole body relaxed; sometimes there are convulsive movements of the whole body; even vomiting and violent perspiration, especially at the stage when consciousness begins to return. The duration of the fainting-fit varies much, as do the degrees of debility and of disturbance of consciousness.

The tendency to fainting-fits varies in different persons; it is not always the weak and delicate women who faint away, but strong and powerfully-built men faint just as frequently— the strongest will-power avails nothing against it. In persons otherwise healthy, strong emotions, also violent pain coming on suddenly, are the most frequent causes; sometimes severe losses of blood occasion serious fainting-fits.

The fainting-fit is brought about by a partial paralysis of the heart and of the large veins in the interior of the body; their walls suddenly become so relaxed that nearly all the blood accumulates in these large pipes, and very little flows through the surface of the body, the muscles, and the brain. From such sudden diminution of blood-supply the brain at once refuses its function: therefore sudden *anæmia of the brain*, resulting from partial paralysis of the heart, is the immediate cause of the group of phenomena termed "fainting-fits." If such a condition lasts several minutes, or perhaps for an hour, it is dangerous — may even be fatal. Generally those who faint soon come to themselves when appropriate remedies are applied — heart and vessels regain their healthy elasticity, the brain regains blood, and therewith consciousness returns. The patient opens his eyes, and soon

realizes his position ; the pale blue-coloured lips again become red, voluntary movements ensue—he raises himself, is
melancholy at first, inclined to vomit, is weak, but soon these
sensations are over and no further consequences are felt.
Fainting-fits very frequently occur in surgical practice, not
only to those about to be, as well as to those who have
been, operated on, but almost more frequently among the
spectators. When the person to be operated on insists
that a relative or friend shall be present at the operation,
I do not object, until the patient is narcotized. He who
does not already know, from experience, whether he can
calmly witness a bloody operation, should not be allowed
to enter the operation-room ; should he collapse at important
moments of the operation, one or more assistants must go
to his help, and the operation is disturbed. In the surgical
clinic, now and then it happens that some one of the young,
vigorous medical students, accustomed to the horrors of the
dissecting-room, faints on seeing the blood streaming from
the living body.

When a person faints the first thing to do is, to lay him
down and keep him with his head lying low until he regains
consciousness. From the changing of the colour in the face
it may at times be seen that some one is about to faint—
by rapidly laying him down possibly this may be prevented.
Sprinkle the face with water, and loosen all articles of clothing
that impede free respiration. Holding *liquor ammoniæ* (hartshorn salts) to the nose is most efficacious ; failing this, rub the
temples with cold water, with vinegar, or with brandy. The use
of much cold water on the head is not advisable. If the fainting
person can swallow, give him some wine, cognac, or coffee,
or ten to fifteen Hoffmann's drops *(spiritus æthereus)* ; or, if
unable to swallow, give an enema with wine. If the fainting

continue a long time, the doctor must be fetched. It is evident that, in fainting-fits occasioned by loss of blood, above everything the bleeding must be stopped.

Very similar to fainting-fits are the conditions produced in *concussion of the brain,* and in so-called *shock,* after severe injuries to the body (for instance, after a fall from a considerable height, extensive contusions, &c.), even without external or internal hæmorrhage.

Very violent, rapid *surcharge of the brain with blood (congestion of the brain)* may also cause unconsciousness, as happens in some diseases of the heart and lungs. In such patients the face and head are then blue-red, bloated, the eyes as if coming out of their sockets, pulse very full, generally slow, the breathing sometimes stertorous. In such cases, high position for the head, cold upon the head, a hot foot-bath, mustard poultices upon the calves of the legs are indicated till the doctor comes and gives further instructions.

In an *apoplectic stroke,* the condition is similar to that last described. As if struck by a heavy blow, the person drops : this is often preceded by a short indisposition, a feeling of weakness, and giddiness in the head. The attack of apoplexy depends, partly upon a change in the distribution of the blood in the hemispheres of the brain coming on suddenly and lasting for some time, and partly, upon the rupture of several small, or of one large, blood-vessel ; the blood agitates the substance of the brain in the vicinity of the rupture, and in this way destroys brain-substance, at the same time that it presses the brain strongly against the inner walls of the cranium. The direct danger to life from the apoplectic attack varies considerably, according to the size of the extravasation of blood, and the part of the brain affected. Occasionally, death supervenes almost instantaneously; in some cases unconsciousness lasts for a few minutes only, in others, several hours; sometimes paralysis of separate parts of the body is the result, or even, paralysis of a moiety of the body. Sometimes movement in the paralyzed parts is fully, or only partially, restored; whilst in other cases, chronic disease (softening of the brain) ensues. The treatment of a person struck by apoplexy must be the same as that of one suffering from severe congestion of the brain.

Convulsions with unconsciousness prevail in *falling sickness (epilepsy)* and are of very many kinds. Some patients feel certain premonitions of the convulsive attacks; but, with many, they come on quite suddenly. After the fit, they have the sensation of having been unconscious—fainting. Many epileptics present a painful appearance during the fit; often it begins with a piercing scream and a twisting movement of the body. The face is distorted, in some pale, in others red, the latter specially when the respiratory muscles are convulsed as well—arms and legs twitch, fists are clenched, the patient foams at the mouth, and the body writhes or is bent together.

There are no means for shortening the duration of such attacks; nothing can be done but to remove everything out of the way by which these unhappy persons can injure themselves; it is best for them to lie on a wide bed, on a carpet, or on cushions quickly laid under them. The popular idea of forcing open the clenched fists is useless, and of no more value is sprinkling, or douching with water, holding stimulants to the nose, or rubbing. None of these things need be done; one can only watch over the movements. Shocking as is the spectacle, and often as the unhappy person appears as if he must die at once, it is very seldom that epileptics die during the fit.

Convulsions of the hysterical may be very similar to those of epileptics, but the hysterical do not lose consciousness—they have pain from contraction of the muscles, though they may not suffer so much as those attacked by traumatic spasm with rigidity *(tetanus)*, lock-jaw *(trismus)*, or hydrophobia *(lyssa)*. In hysterical convulsions the sufferer at times would seem to be almost dying, but death from this cause scarcely ever happens. In such cases, remedies applied to the nose, sprinkling with water, and rubbing the temples with eau-de-Cologne may be successful.

By *colic, painful contractions of the abdominal organs* is understood. There is intestinal colic, gall-stone colic, and colic from stone in the kidney. The pain (generally described as "cramp," but in reality only the result of the cramp) may be so violent as to cause fainting fits, cold sweats, and a feeling of approaching death. These colic-pains are best removed by heat. Warm, even moderately hot, compresses on the abdomen alleviate, and are most effectual. Mustard poultices on the abdomen (see p. 116) sometimes give relief. The injection under the skin (sub-cutaneous injection) of a solution of morphia is most certain in its operation, but it must not be done except by a nurse experienced in so doing, and then only when ordered by the doctor.

What is understood as *spasm of the stomach* is very often gall-stone colic, but there are painful contractions of the stomach which arise from ulcers, and principally in anæmic young girls. When such cases are combined with gastric hæmorrhages and blood-vomiting, the application of warmth might prove dangerous : then the nurse must do nothing independently. The assistance to be given in vomiting (also a species of morbid movement of the stomach) will be found described on page 100, on "Administering Emetics."

Respiratory spasms are the most distressing. Those attacked raise themselves upright instinctively, as they feel in danger of suffocation.

Spasm of the glottis, which may attack healthy but sensitive persons, is caused, at times, by choking (from "having swallowed the wrong way"), or by saliva running into the larynx, or by loud talking and coughing with the vocal chords already irritated by catarrh. Those who have suffered from spasm of the glottis frequently feel beforehand that the spasm is likely to come, and then they must keep very quiet —must neither speak, nor laugh, nor clear the throat, but must take slow, short inspirations until they feel that they can again breathe freely; even then, for a time, rest is necessary. If the spasm come, the sufferer appears as if he would be choked; with every effort to draw breath little air passes the glottis, sometimes with a sibilant sound; anxiety increases with the length of the spasm: this rarely leads to actual suffocation, but gradually lessens. The person attacked must keep very quiet: sprinkling with water, or rubbing the forehead and temples will possibly shorten the spasm, but is unpleasant to many in this state. Such a spasm seldom lasts in its full violence more than from one to two minutes; it easily returns, if perfect quiet is not maintained for some time afterwards.

The spasm of the glottis in little children, a symptom of rickets *(rhachitis)* which may lead to suffocation, deserves special attention. Quickly putting the forefinger far into the child's mouth, pressing the root of the tongue downwards and forwards (against the teeth), will usually cause the choking fit rapidly to pass away. Judicious medical treatment is necessary to obtain a cure quickly.

Asthma, the *asthmatic paroxysm*, is another species of

spasm of the respiration ; it mostly comes on only in certain diseases of the heart and lungs. During the attack, which is combined with fearful difficulty in breathing, the sufferers endure, as it were, mortal anguish, and always want something done for them. Loosen all articles of clothing producing restraint, convey fresh air to the patient quickly by opening the window, and leave him free to determine the position of his body. Very *strong* coffee, or, from time to time, small portions of fruit-ice, rubbing the chest with oil of turpentine, mustard poultices on the chest, calves of the legs, and arms, hot hand and foot baths, must be given ; in one case, one remedy—in another, another most avails. From experience, persons generally find for themselves what most easily provokes the attack. Some patients are free from attacks during the night only when a light is burning ; others must have the door open to the adjoining room. For all asthmatic persons, residence in places with dry, pure air is advisable, with avoidance of wind, of dusty and smoky atmosphere, and moderation in living and in sleeping ; whilst, for patients with catarrhs, or with consumptive tendencies, damp, warm air is more beneficial.

Finally, another form of respiratory spasm must be mentioned—the " *hiccough* " *(singultus)* or "hiccup ;" it is a contraction of the diaphragm, occurring at intervals by fits and starts, sometimes caused by long violent laughter, and also by gastric irritation. Children get it very easily ; they must be kept quiet, and water, or sugared water, given them to drink ; a few drops of vinegar on a teaspoonful of crushed sugar often prove useful. Long retention of the breath avails most. In continuous hiccoughing for days in succession, resort must be had to medicine, which the doctor must prescribe.

The violent paroxysmal pains attacking the sensory, especially the facial nerves, are called *Neuralgia* (tic). It is one of the most frightful sufferings man can endure. Imagine the most furious pains in all the teeth on one side of the face, coming on suddenly like lightning, and an idea may be formed of the terrible nature of neuralgia—it is very difficult to cure; the unfortunate sufferer may obtain alleviation by strongly pressing the whole hand upon the painful side of the face. If no relief can be obtained, then cutting through the nerves may be of use, and generally the operation is not dangerous.

OBSERVATION AND CARE OF THE MENTALLY DISEASED.

Nearly every acute disease may occasionally be accompanied by mental derangement; that this occurs (though not in every case) simultaneously with high fever, and with disease from blood-poisoning, is stated at p. 191. But many chronic affections of some part or parts of the body—from the intimate connection existing between all its organs and functions—can so irritate the sections of the brain in which mental functions have their seat as to cause mental diseases. Finally, these parts of the brain can of themselves become diseased. Thus the immediate causes of mental diseases lie in the derangement of the functions of the brain; yet by no means has the derangement always begun in the brain, but it is often initiated by disease of some one or other, and possibly remote-lying, organ. That the recognition of such close relationship is of the highest importance to the treatment, is self-evident.

Often enough, disturbances of the motor and of the sensory functions are combined with derangement of the mental functions. Partial paralyses of tongue and lip movements necessary to speech most frequently occur, and not seldom, epilepsy also (falling sickness, page 248) ; lastly, paralysis of the limbs may be also combined with enfeeblement of the mental functions (with imbecility, idiocy).

In institutions for the insane, naturally only such female attendants should be appointed as have been specially trained for this branch of sick-nursing. But every nurse, although she may not wish to become specially a nurse to the insane, and *every educated woman*, should have an unprejudiced idea of the nature of diseases of the mind, because many dismal stories about such patients are circulated, which, unfortunately, here and there still lead to ill-treatment of these unhappy people.

For the instruction of all who may have the care of persons suffering mentally, I will here give the most

IMPORTANT RULES ON THE OBSERVATION OF THE MENTALLY DISEASED,

in the words of Dr Ewald Hecker, from his admirable essay on " Nursing the Mentally Diseased," in the " Weimar Vade-Mecum for Nurses, 1880."

Dr Ewald Hecker says :—

" The mental (psychical) symptoms of diseases of the mind in their vast totality may be divided into three groups ; these, however, are not actually separated from each other, but occur in most varied combinations. The first group consists of symptoms which indicate derangement of the imagination *(ideal conceptions)* ; the second, all phenomena denoting a repression and abatement *(depression)* of intellectual power ; whilst the third embraces the conditions of excitation of the mental function *(conditions of psychical excitement)*.

"I.—DERANGEMENTS OF THE IMAGINATION

(THE IDEAL CONCEPTIONS AND THE IMAGINATIVE FACULTY).

"The derangement of ideas called *insanity* may be combined with the conditions of excitement or of depression; it may, however, come on when the emotional life is in a state of complete tranquillity and seeming equilibrium. As the characters of the fundamental dispositions differ, so hallucinations bear a different stamp. Whilst the melancholic, depressed patient believes he is persecuted, calumniated, ruined pecuniarily, burdened with sin, mistaken as to his salvation, the cheerfully disposed imagines himself in possession of untold wealth, high titles and orders, the greatest wisdom, great bodily strength, &c., &c.

"The source whence the subjects of the hallucinatory ideas are drawn is, in such cases, the morbidly excited condition of the mind. Another frequent cause of insanity arises from *delusions of the senses* (hallucinations and illusions). All the senses—feeling, smelling, tasting, seeing, and hearing—can be thus affected, and this arises from irritation of the terminal organs (located in the brain) of the affected nerves. In accordance with the habit of health, the patient still assigns to external origin those excitations known to his perceptions hitherto as arising only from without, and so he now believes, when suffering from hallucinations of sensation, that his body is tortured, beaten, cut in pieces, treated with electric currents, persecuted by inimical powers, or is the seat of a strange being; or, in hallucinations of smell and taste, that his food is poisoned, the air around him is pestilential; that he himself is thoroughly infected with a loathsome disease; the sufferer from optical hallucination sees figures and situations before him, now pleasant, now unpleasant, flattering or threatening ; whilst hallucinations of hearing are of the most manifold character—now the words are isolated (mostly invective), now he hears whole sentences or speeches which answer his questions, or, without his suggestion, perplex him continually, and incite him to certain, often unnatural, actions. Occasionally, hallucinations of different senses combine together in one, all the more delusive—he sees the murderer who breaks in upon him, and he hears the crack of the shot that wounds him ; and so on.

"In the so-called *illusions* it is a question of misconstruction, an interpretation of an actual perception of the senses in a contrary sense. The patient believes those around him not to be the persons that they are : a relative is suddenly metamorphosed into the devil ; a cloud becomes the form of God ; he reads in the newspaper sentences wholly different from what they are ; he makes words out of, and gives meaning to, the chirping of the birds, the whistling of the wind, the ringing of bells, or conversations held at too great a distance to be understood.

"In most cases it is not difficult to prove the existence of hallucinations, but sometimes patients do not express any opinion upon the subject—indeed, they deny them. Thus it is of great importance to determine, from the whole demeanour of the patient, whether he is hallucinated or not. The *symptoms* which, with tolerable certainty, indicate a state of hallucination are : breathless listening at a fixed spot, a persistent look in one direction, stopping the ears, hiding the face from view, an invective word spoken aside suddenly, stopping-up cracks in the walls and floors, at the same time intently listening, &c. Also, by coining new words, by temporary dumbness, and by refusal of food, the state of hallucination is often recognised. It is very important to know, that hallucinations are the most frequent source of wholly unexpected *acts of violence or of suicide*, in striking contrast to the character of the patient.

"There are isolated cases in which *hallucinations occur in persons otherwise mentally sound*, and who are conscious of the morbidity of these phenomena. Similar to this is a form of mental disease, not rare, in which certain ideas (mostly of fear that he has done, or is capable of doing, something wrong) are continually obtruded upon the otherwise intelligent patient. He realises that these have no foundation in fact, but is unable to rid himself of these uncontrollable and frequently-returning thoughts and ideas. As a result, anxious fears of touching persons or objects are created, from which patients fall into a really deplorable condition.

" *The fixed idea*, although it frequently appears as a symptom of mental disease, is yet by no means an absolutely necessary characteristic indication of it. There is a considerable number of mentally diseased persons who are not dominated by any hallucination, but whose disease, without being in any way less severe or dangerous, declares itself only by the conditions of depression or of excitation now to be discussed.

"II.—CONDITIONS OF MENTAL DEPRESSION.

"The symptoms of repression or of depression at first affect the emotional vitality of the patient, and find their expression in so-called *melancholy*. Incidental to the primary stage of almost all mental diseases, melancholy essentially denotes, as the term itself implies even to the unlearned, a sorrowful, low-spirited condition of the mind, with suspension of the emotional flow of the feelings and ideas which would otherwise exist. Only the gnawing sensations of remorse, of pain, of grief, of jealousy, of unrest and anxiety, affect the patient, whilst ideas, which would otherwise produce satisfaction and pleasure, glide by leaving no trace, or are changed into their very opposite character. Illusionary ideas, in keeping with this low-spirited condition, are often present, but sometimes are altogether wanting. In every case the patient's thinking power is paralyzed in a greater or less degree—he can follow his ordinary occupation only with difficulty, or not at all; his thoughts move along laboriously, almost exclusively revolving around the one sore point. The will-power is also relaxed. He sits inactive, unsympathetic, often not unlike an idiot in whom the repression of all mental functions depends upon the extinction of the mental powers. But when the extremely painful symptom of *fear*—often coming in paroxysms—is combined with the melancholy, the patient is in a state of great bodily unrest. Loudly lamenting and screaming, wringing his hands, tearing his clothes, scratching his skin sore, clinging to every one who comes near him, he runs restlessly round day and night, a picture of the greatest terror and mental agony. With all persons suffering from melancholy, attempts at suicide must be anticipated; sometimes these are threatened beforehand by the patient, but suicide is often committed without notice, and with much calculation and cunning. *Refusal of food*, as a means of suicide, is not seldom practised, but this originates also in other forms of hallucination, in particular, from hallucinations combined with the insane delusion of being poisoned.

"Standing apparently in contradiction to the character of melancholia, and therefore all the more noticeable, are *the*

R

acts of violence perpetrated upon others by such patients. In some cases, in moments of the greatest terror, but in others, apparently with cool calculation, influenced by melancholic hallucinations and delusions, many of these sufferers have slain their dearest relatives.

"To the conditions which reveal mental depression belong *imbecility, craziness*, and *idiocy*, and these constitute the final stage of the different forms of mental disease. Destitute of every deeper sensation and emotional excitement, without regulated ideal conceptions, the patients live on without thought for the morrow, often swayed by abnormal impulses (acquisitiveness, kleptomania, &c.).

"III.—CONDITIONS OF EXCITATION.

"The conditions of mental excitement in their most developed degree constitute *madness*—then the patient wanders about chattering senselessly, screaming and singing with a lively, rapid flight of ideas, now joyous, now passionately excited, destroying his clothes and effects, and violently attacking all around him. He is often governed by hallucinations and rapidly changing illusionary ideas, and it is in the nature of the disease that these readily bear the character of raving madness.

"But the excitation does not always increase to so high a degree. There are forms of the disease the characteristic feature of which is, a peculiar excitation of the brain functions, such as is present in a person slightly intoxicated. The patients appear more clever, wittily and intellectually, than in days of health, and they state quite artlessly, and with much ingenuity when necessary, their reasons for the very numerous foolish actions, now childish, now exaggerated, now criminal, which they commit. In a way, easy of explanation, they are often regarded by the uninitiated as in full health ; and, from this reason, the greatest mischief ensues. Their disease often appears to be a fault of character, a moral defect, similar to that of confounding one thing with another,—this form has been designated *moral insanity.* They are real experts in this : they compromise their families by indelicate communications ; they playfully jest with their surrounding company ; they play tricks, sow discord, and finally put all around them into confusion, and a state of mental excitement. With a gift of observation sharpened by disease they understand, in a masterly way, how to detect and utilise the faults and weaknesses of others, and it is because of this that nursing such patients is extremely worrying and difficult, and can only be endured to the end by *the nurse who never forgets* that she has actually to do with one mentally diseased.

"It is with these patients that a *periodic interchange* of melancholy and excitement frequently appears in constantly recurring attacks, separated from each other by a stage of apparent health."

GENERAL DIRECTIONS FOR NURSING THE MENTALLY DISEASED.

It cannot be too widely known that *diseases of the mind are also diseases of the body*, and, like all other diseases, depend upon changes in the bodily organs, and especially in the brain.

It is not easy to make this quite clear to the uninitiated, for it is too much the custom to consider the soul of a man as something almost independent of his body, and to regard and form a judgment of it by itself. But the nurse of the mentally-diseased must always realise that *she has to do with persons pitiably and bodily sick*, though she may see very little illness in them besides the raving. On this point, Dr E. Hecker says :—

"The nurse will then endure more easily the many disagreeables that arise in nursing patients mentally ill; she will not lose her equanimity or her gentleness when she hears them say, and sees them do, things which wound her feelings.   Yes, even when personally ridiculed, affronted, calumniated, or even violently attacked by them with seemingly calculated malice, feelings, not of displeasure and anger, but of compassion will be excited in her.   She well knows that the *person mentally sick is in no way responsible* for what he does and says, inasmuch as he lies under the ban of the disease, which is able to produce a complete subversion of his character, as well as of the whole method of his thinking and acting.   The man, previously sensitive morally, who loathed every vulgarity from the bottom of his heart, has become indecorous; he who always controlled himself has become unrestrained, the cowardly, courageous; he who was full of tact, becomes regardless, and so on.   In this state, no preaching, no scolding, nor reasoning is of any use—*insanity is not to be dispelled by argument.*   The disorder of the mind, with all its individual symptoms, depends upon disease of the brain, and is just as little removed, or even diminished, by talking to the patient, by disputing or discussing with him, as paralysis or a bodily pain would be; on the contrary, to contend with him about his delusions operates like a continual pulling and tearing at a wound; the diseased portion of the brain becomes still more excited, with the result that the fixed ideas, the sensorial illusions, the anxiety of mind, take all the firmer hold, and increase.

"Which, then, is the correct treatment of the delusions of

the patient?" continues Dr Hecker. "Ought the nurse to consent to, and confirm the patient therein? By no means. Shall she repulse him abruptly and harshly, laugh at, or deride him? Just as little. But there is a certain kind way, adapted to the various circumstances and characters, which may, in individual cases, clothe itself in a light jest, a hint that the future will teach him better, or a reference to medical opinion, and other means, left to the fine tact of the nurse, by which to avoid agreement with his delusions without irritating him. In every case she must so act that he feels her sympathy is that of the heart. By small harmless services which she can render, by readily consenting to innocent wishes, and so on, she will soon know how to give further practical proof, at the right time, of her sympathy, and thereby gain his confidence. From the irresponsible condition of her ward she must never allow herself to be betrayed into forgetting the social respect due to his position, for many patients, and precisely those to whom least credit for it would be given, preserve very refined, delicate feelings, and an active perception of everything that is called social tact, even through the severest stages of the disease.

" It is very wrong and unjust for a nurse to seek to gain the patient's confidence by being untruthful, to pacify him by misrepresentation, to put him in a good humour by promises not seriously meant. By such means she may be easily placed in an utterly untenable position, and awaken directly the mistrust of the patient. On the other hand, she is not prevented from exercising a certain *reticence* and *caution* in *speech*, and from withholding such communications as would be likely to excite and disturb him. Besides, she must not forget that a person with a diseased mind takes offence at many things, and is injured by many words, which, to the

sane, appear harmless; just as a person with a sick hand
feels as pain every gentle touch, although in health he would
be scarcely conscious of it.   Exactly that which is beneficial to
the sane, often evokes in the insane the contrary sensation.
Lively stories depress and irritate the melancholic still more;
he who is dominated by mania, marked by the delusion of
persecution, sees derision and scorn in harmless expressions
of friendship, and the nurse often finds difficulty in using the
right tone and terms in conversing with him.   Keen observation and fine tact will help her in these circumstances—there
are times when she will know how, judiciously, to be quite
silent.

"Of course she must take care that the irritating influences
described do not approach him in other ways, or possibly, from
reading injudiciously chosen.   That narratives of dreadful
purport, descriptions of harrowing scenes of human misery,
tales of murder and suicide, are not adapted to melancholic,
anxious persons, is evident, and therefore the nurse will do
well to place no book in the hands of her charge which she
has not herself read.   She will find further, that many juve-
nile and popular stories, seemingly harmless, are not suitable
to some sick persons.   The same holds good of books of
religious tendency, which have caused much harm; and, for
very many patients, even of the daily papers, which are best
wholly banished from the sick-room.   Finally, some must be
kept from reading altogether, as they interpret in a contrary
sense, and apply to themselves all that they read.

"Similarly, *letters* and *visits* may have a prejudicial effect,
and therefore the nurse must not deliver any written document
to the patient until it has been examined by the doctor, nor
admit any visitor without his express permission.   In most
cases, especially at the commencement of mental disease, to

remove the patient from all that previously surrounded him is the primary duty, for the morbid thoughts and feelings, in their gradual beginning, are almost always closely connected with the various objects, localities, or persons familiar to the patient, and in such a way that, by the sight of them, the remembrance is revived. Thus it is explicable that, even in cases in which morbid ideas and feelings, such as anger, distrust, jealousy, hate towards relatives, have no existence, and where complete security from thoughtless communications and exciting conversations on the part of the visitors would be ensured, still an unseasonable visit of relatives can cause the greatest injury to the patient."

I am indebted to the kindness of my colleague, Dr J. von Mundy, for yet further Instructions for Nurses, as follow :—

"Although it is a fact that many thousands of the insane, idiots, cretins, &c., are not placed in special institutions, but live in their own families, or with strangers who are paid for taking care of them, yet it is beyond question that the majority enjoy no care adequate to the scientific and humane principles now current; consequently, many of these unhappy people are greatly neglected, and not seldom even ill-treated. The nursing necessary for the sick, or for the wounded, is different, in infinitely many respects, from that necessary for the insane. For with the former, as a rule, one has to do with a person in full possession of his faculties, who thinks and judges sanely, whose will is free, and who, according to his strength, can comply with the wishes or advice of his nurse ; for the most part he is dangerous neither to himself nor to others. Those about him can assist the nurse in her duties, or take her place in her absence.

"In nursing the insane the contrary is the case—the family

and the domestics are embarrassed and anxious; the symptoms of the disease often become most violent when a member of the family enters the sick-room and desires to communicate with him. Many insane persons are dangerous to the public or to themselves, and must be watched without intermission— must *never* be left alone. They are generally insensible to every exhortation, every attempt to advise, every word of comfort. Often dirty, they must be constantly withheld and protected from evil habits injurious to them, and from un-mannerliness. Very frequently they refuse all food and drink, specially medicines. They rave, become rampant, and destroy everything within reach ; howl, strike, scream, cry, pray, swear, and consider themselves persecuted, condemned, poisoned ; throw themselves upon the floor, strip naked and tear their clothes to shreds. Or again, they are very dejected, squat upon the floor, shut their eyes, and wrap up their heads ; declare that they hear voices and see persons, who speak to them, and whom they answer ; they talk to themselves, often very vivaciously, until they become hoarse, fancy certain of their limbs are wanting, or that they have been exchanged, frozen, or burnt, or put on the reverse way ; deem themselves to be extraordinary persons, or gods, devils, kings, &c. ; as very rich or very poor, as deceived, and unjustly deprived of their freedom, &c.

"It is precisely the nurses, male and female, who will be regarded by such patients with suspicion, will be hated, calumniated, abused, deceived by falsehoods, at times attacked violently. Other lunatics appear, indeed, mentally quite sane and vigorous, so long as the sore point of their diseased power of imagination is not touched upon by themselves or by others. This often declares itself as a definite delusion (fixed idea), disappearing only with death. Most patients—of the various forms of whose disease only some disease-phenomena in general have here been briefly given, by way of example—are also bodily ill, and require special bodily care. For instance, they suffer from cramps, from frequent fits of unconsciousness, from partial or general paralysis of the limbs—often cannot walk along without support; they feel, hear, taste, smell, see, badly or incorrectly, have new or old wounds and ulcers on the body, are unable to discharge their evacuations without help, or to dress and undress themselves, &c.

" From this cursory sketch it is evident that there is great difference between ordinary sick-nursing and the nursing of, and attendance upon, the insane. But many things must be added which will considerably increase this difference, making the care of the mentally diseased so difficult a task that very particular, and specially long, experience is necessary for its happy solution. In the course of years cases of disease occur in most families, about which the relatives and domestics, even friends and visitors of the house, hear something, participate in the nursing, get accustomed to these occurrences, and thus gather experience, though it may be very imperfect. For classes less comfortably circumstanced, it is instructive occasionally to stay in hospitals or to visit them, and to observe the directions therein followed relative to many things serviceable in the domestic nursing of the sick.

"It is otherwise with mental diseases, because, in comparison with other sicknesses, the sufferers from them are much less numerous, and they remain only for a short time in their families to be nursed ; and because also, admission to institutions for the insane is, with justice, much restricted. Consequently those who desire to fit themselves for the care of the insane have, as have also the general public, little opportunity, either in or out of these institutions, to learn this difficult branch of sick-nursing.

" We shall now briefly consider that part of the Care of the Insane that is closely related to Domestic or Family Nursing.

" On a nurse being summoned to take charge of an insane patient in his house, her first duty is, even before entering the sick-room, to ascertain, briefly but certainly, his condition at the time. The doctor can tell her best; in his absence, the nurse must get the information from the member of the family who seems most composed, and in so doing, where possible, keep aloof from all others present so as not to be continually interrupted and perplexed by their observations. Whether the case is of long standing (chronic), or newly arisen (acute), will soon be proved; and which of the forms of disease we have already described is present, will quickly be indicated—whether that in which chiefly the imaginative faculty is deranged, or that in which the condition of mental depression or of mental excitation prevails; and further, what other bodily weaknesses may be combined therewith.

" It is exceptional for a nurse to be called hastily into a house to nurse a quiet lunatic. As a rule, her assistance will be required in new (acute) cases, in which the patient's condition of excitation has reached so high a degree that no one in the house any longer knows what to advise or how to help.

" The unhappy madman who, in such cases, often very violently raves, screams, menaces everybody, and destroys everything, creates such a fear in those about him, and in his neighbourhood, that no one will venture to approach him— his perplexed and helpless family leave him mostly to himself, and confine him to the house.

" Here the nurse's first duty on arrival is, by courage, coolness, patience, and discreet instructions, to inspire those around the patient with the like spirit, and to rouse them to rational, quiet action.

"In such a case the nurse must have all furniture and movable objects quickly cleared from one of the rooms, and the windows must be firmly protected to a height beyond the patient's reach with mattresses, beds, laths taken from bedsteads or from the backs of wardrobes, or they must be screened off.

" In the empty room, a mattress or a large palliasse must be laid upon the floor. Then the nurse, as noiselessly as possible, must bore a small hole in an angle of the panel of the door of the room in which is the maniac; through this little hole (not the keyhole, as this the insane always watch suspiciously) she must observe him. When quiet, she may open the door, to induce him to leave the room. If he still raves, she must take care lest he hold, ready or hidden, some dangerous article (weapon, pointed instrument, glass, piece of wood, iron, steel, etc.) wherewith to inflict injury upon the person entering. Among a *hundred* cases this will hardly occur *once;* but it is prudent to act *always* as if the danger existed. If this is the case, she must not open the door of the maniac's room until she has the assistance of, at least, two resolute men. Opening the door cautiously, they must not enter, but must endeavour to persuade him to leave the room voluntarily. Whilst this is taking place, the nurse must keep herself at a *safe* distance, until, should he refuse to leave the room, the men succeed in approaching him so as to take away the dangerous article, and in laying hold of him with strength, but with forbearance. The nurse then orders him to be brought into the prepared empty room, and locks him in. Through the small hole (peep-hole), bored in the door of this room also, she will be able in a short time to see if the unhappy madman gradually becomes more quiet, and lays himself to rest upon the improvised couch on the floor. For, after the state of fearful excitement and violent exertion of all his muscles in his phrensied movements, *a condition of relaxation and exhaustion*, even fainting, naturally follows. If fainting occurs, the nurse must immediately apply the remedies prescribed on p. 246. To calm the wearied maniac, and to prevent rapid relapse, the best remedy is, to *refresh him with food*. For this the nurse must make preparations, so that, in the intervals of quiet, he may partake of warm tea or coffee, or strong beef-tea, or solid foods and cooling drinks. Even wine, rum, or cognac, in small quantities, may be given to such patients, when not suffering severely from determination of blood to the brain. For the insane, with serous or too little blood, spirituous drinks given moderately are positively necessary.

" When called to assist in the frequently occurring attacks of 'drunkard's madness' *(delirium tremens)*, the treatment by the nurse must be similar to that just described. These persons act like raving madmen, but complain, at the same time, of heavy pressure upon the chest, with constriction of the throat, and fancy they are persecuted by mice, rats, and other little creatures, or imagine they see fire, smoke, lightning, etc. In such cases, according to medical prescription, wet packings, or cold affusions and cold wet compresses on the head and chest, must be applied; subsequently, very strengthening restoratives and spirituous drinks may be given. To guard against possible danger, it is best to confine the patient in a small room, converted into a madman's cell as already described.

" The same treatment holds good for such *epileptics* (those who suffer from 'falling sickness') as are often attacked by severe fits, with violent convulsions, throwing the body to and fro (see p. 248). A couch, of several mattresses arranged side by side on the floor is, in such cases, particularly necessary, because, if the patient were to fall out of bed he could easily injure himself against its angular edges, or against other articles standing in the room.

"With persons suffering from *delirium tremens* (when the help of strong men is often required), and during the fits of epileptics, the nurse must not leave the room; she must have a chair and table placed for herself at some distance from her charge. If, after such fits, a bath is ordered by the doctor, she must supervise the degree of temperature and length of time in the bath with great exactness, and, under no circumstances, must she leave the room during the bath.

"The question must now be considered whether, in the paroxysms of the insane, the strait-jacket or other means of restraint may be used by the nurse. This apparatus, hated by all patients, a hindrance to the circulation of the blood and the breathing, has been abolished by law in Great Britain for more than forty years, although there are more than 100,000 insane cared for in her institutions. The padded cells are used, and, in special cases, the strait-jacket is replaced by temporarily holding the patient firmly by the powerful hands of numerous well-trained, well-paid keepers. Therefore, this method of treatment is mostly described by the expression, 'No restraint' (without constraint).

"For the insane, *when privately nursed*, absolutely the same principles should prevail; and though it is to be regretted that, in very many European and American establishments, this is not yet the case, still, in private nursing in wealthy families the strait-jacket should all the more readily be abolished, because, with single patients, a sufficient staff of keepers, day and night, is more easily secured than is possible in asylums, with their limited State and Provincial funds, and the many patients crowded into them. The conditions of excitement of the

mentally diseased are always much more transient when
abnormal movements are mastered by the hand than when
by the strait-jacket, as the latter very materially increases the
raving fit.   By the violent movements of the patient in this
apparatus, excoriations and other injuries, as well as inflammations of the organs of the chest and abdomen, are
occasioned.   There have been cases where, from the severe
constriction, apoplectic fits ending in sudden death have
occurred.   For such reasons the use of the strait-jacket, the
securing by cords, by linen, or by straps, must be urgently
condemned.   Of course, in firmly holding the raving patient
the keeper must grasp him with care and skill; the male
assistant to the nurse must *never* be permitted to use his feet
or his knees, possibly pressing them upon the chest or upon
the abdomen of the patient, who has been flung upon the
floor.   One keeper should hold his arms or his hands, a
second and a third his feet, and, in serious cases, a fourth
his head and shoulders.   A certain practice is necessary for
this duty, and by attendance upon persons when in the raving
fits under chloroform it may be acquired.   Experience teaches
that, as a rule, the patient becomes composed in a comparatively short time, whilst, if the strait-jacket be used,
he may rave for hours, even days, refusing all nourishment, and, from his continuous screaming, keeping the house
and the whole neighbourhood in a state of anxiety and
dread.   More and more the principle of 'No restraint' is
spreading, although slowly.   In the Lower Austrian Provincial
Asylum for the Insane, the Director, Professor Dr Schlager,
has proved, that the use of the strait-jacket is unnecessary,
and, in dispensing with it, excellent results have followed.

" To prevent *attempts at flight*, which occur frequently in the
condition of excitement of madmen, *and are not rare in*

other forms of insanity, an uninterrupted supervision must be maintained.   For this reason it cannot too often be repeated, the nurse must *never* leave the mentally-diseased person alone: when she must be absent she must be replaced by a thoroughly reliable person.

"With all insane persons suffering from depression *(melancholy)* the nurse's task is very limited.   When this is not accompanied by conditions of excitement and disturbance of the imaginative faculty, as may be the case, the nursing is almost entirely physical.   On the whole, the nurse must not expect much success from her influence, mental and moral, upon the mentally-diseased patient entrusted to her care.   *The silent discharge of her duty, and the caring for nothing but the patient* are the supreme principles which must guide her in the treatment of *every form of insanity.*   Now whether the madman, tormented by every kind of fear and fancying himself persecuted, howls, cries, and complains, or whether he sinks gloomily into sombre reveries, or excitedly strides unceasingly about the room, or even squats upon the ground and will not permit himself to be raised, or continually repeats words singly, makes speeches, asks questions, or is always silent, and, even when he can speak, cannot be induced to say either 'Yes' or 'No'—still the quiet, earnest, silent discharge of her duties is most successful.

"*Apparently* to regard the insane as 'sane,' and also *only apparently* to treat him as such, is still further most important counsel to the nurse.   In loss of memory, the effort to remember renders him helpless and perplexed, particularly in new cases.   Here, the nurse must think for the patient, and gently but quickly come to his help in everything.   For instance, the generally, although not completely, incapacitated

person, hears the carriage rolling rapidly behind him but
makes no attempt to get out of its way, because he has lost
the recollection of the danger to which he is thereby exposed.
At times he imagines the nurse is his brother, or his deceased
sister.   To the insane, many things appear, often everything,
enigmatical, mysterious, inexplicable.   Possibly the nurse herself seems to him a most fantastic creature.

"With patients suffering from suicidal, or mutilating mania,
it is a matter of course to double the number of the attendants, and to give no credit to the most subtle representations,
in which these patients are very inventive.   The clothing, and
the beds of such—generally of all—insane persons must be
very often searched; inflammable articles, naked lights, everything by which injury could be caused, must be removed from
the room.   The medicines the nurse must always keep under
lock and key.   The patients must not be permitted to shave
unless they use the so-called safety razor; when clean of
habit, the beard may be allowed to grow; in other cases, they
must be shaved by skilful attendants experienced in the use of
the razor.

"It is a very difficult task for the nurse when the insane
refuse food, and persist in so doing.   'Persuasion' has no
effect, for the refusal is a special form of mental malady depending upon some disease of the imaginative faculty (illusion
or hallucination); or upon a state of depression, or even
upon an impulse to self-destruction.   Many such patients
fancy that they are too poor to pay for their food; others, that
their food and drink are poisoned (delusions of smell and taste
—persecution mania); some fear that they have no power to
chew, or to swallow food, &c.   In such cases, by seeming
absence of mind and negligence, the nurse will generally succeed in exciting the patient to eat by removing the food and

drink from the table and placing them in a corner of the room, or in a cupboard. Of course, the food so placed must be renewed from time to time. Frequently the patient is so greatly tormented by hunger that, if the nurse appears to take no notice of him, he will hastily go to the food and satisfy himself. Very seldom, or never, will she be able to induce patients to take food by talking to them, or by personally eating. Often unfortunately, in such cases, artificial feeding by means of the œsophageal tube introduced into the stomach, must be resorted to. At first this is done by the doctor, but a skilful nurse will easily learn how to do it.

" In all *derangements of the imaginative faculty*, especially in delusions of the senses, a quiet, seemingly indifferent, deportment is the best method. With the insane, every contradiction, all advice, falls upon barren soil. For instance, an unfortunate person who imagines that he is a king and has decorated himself with many orders, is not to be convinced of his delusion, even if one clothes him in rags. A patient who believes himself possessed by Satan is just as little to be quieted by prayers said over him.

" Accordingly, when the mental and moral relations of the nurse to the insane can only consist in close supervision and observation, she must apply her energies all the more to the physical care of the patients, and specially to keeping them clean. As a rule, the insane, idiots, paralyzed, and epileptics are very dirty, and here is an ample and exceedingly important field of activity for her. She must wash, comb, and dress them (which such patients submit to very unwillingly), cut their nails (so that they shall not scratch open their skin), wipe the nose, feed them, assist and accustom them like children to regular evacuations, change their linen when necessary, often several times daily, lead them, accompany them in their drives, and, in brief, treat them quite as great overgrown children. That is a task in which, by means of one of their greatest virtues—courageous self-sacrifice, ready devotion—women fulfil, in the highest degree, the service one person is able to render to another. To die for any one, only a brief moment of rapid resolution is necessary; but, through long years to suffer patiently with, and from, an unfortunate person, that is in truth great and noble, all the more great and noble because scarcely acknowledged—for the reward lies only in the self-consciousness of the most difficult duty faithfully fulfilled.

"Another point in the care of the insane, relating to the important questions of their tranquillity and occupation, still remains to be considered. In this, as in everything connected with the dietetic treatment, in each separate case the nurse must receive her instructions from the doctor. If the form of insanity is such that absolute rest of body and mind is ordered, or if the patient has no longer sufficient power for any work, or if he is completely insensible to all amusement, or is the same hurtful in his diseased state, then, of course, activity must cease, and to secure absolute rest as ordered is the duty of the nurse.

"In other cases she must endeavour, as directed by the doctor, to ascertain the patient's favourite occupations, to promote and superintend them, to stimulate him to work, and to assist and take part in his varied occupations and amusements (reading, writing, drawing, sewing, embroidery, card-playing, &c.). In all modern institutions for the insane it is now an established principle, that they shall be combined with workshops, farms, and gardens, so that, not only the insane able to work, and fit for it, shall be occupied in the workshop, the field, the garden, and the kitchen, but also, that patients from the higher classes may occupy themselves as much as possible with pen, pencil, or chisel, or may amuse themselves with music, singing, or dancing.

"Occupation and work are the most important means for strengthening the body, and for diverting the mind of the insane. And although previously remarked, that the care of the insane in families still leaves much to be desired, yet it must be stated that, in one special district in Belgium—Gheel—for a very long time past, in thirteen villages and one town within a circuit of nine miles, many hundreds of persons, suffering from all forms of insanity (at present over 1200), live at liberty, without restraint, and are taken care of in the families of the inhabitants.* Even here, besides being cared for in the family with free unconstrained treatment, work,—in the meadow, in the field, in the stable, in the workshops as well as in other occupations (domestic service, nursemaid's work, kitchen work, washing, sewing, lace-making, &c.)—forms the most important remedy for promoting improvement or recovery of patients. Excepting cases springing from heredity, or from casually caused diseases of the body, it is mostly grief, misery, anxiety, losses, bad habits, such as drunkenness, gambling, and other passions, that give the first impulse towards mental and bodily diseases, and in which, regular work, regular passing of the day with moderate diversion, with a degree of physical fatigue necessary to healthy sleep, help most to lead the mind—driven out of its ordinary course, swaying to and fro—gradually back again into the right track, and into normal action."

* None are admitted into the Colony of Gheel having dangerous tendencies, suicidal or homicidal. In all cases, duly attested medical certificates are necessary, and official permission to enter the colony must be obtained from the Burgomaster. The charges and location vary, according to the requirements of the patients.—TR.

CHAPTER VIII.

AID IN ACCIDENTS.

IT has already been distinctly stated, that the nurse must not play the doctor, but has only to carry out the doctor's instructions. Yet sudden emergencies and accidents do occur, in which, partly impelled by her own desire to help, and partly urged thereto by others, she cannot refrain from quickly acting independently. As, from her hospital training, she is accustomed to see cases of accident and drowning, so she will do what is necessary more quietly, and with more certainty than others who, from fright, do not overcome their lamentations and crying. Should she be unable personally to assist, still, from her practical experience, she will be able to prevent anything injudicious from being done before the doctor arrives.

INJURIES.

Burns cause, either a redness of the skin, or the formation of blisters, or a cutaneous eschar.*   If the clothes of a person have caught fire, then immediately throw him down, put a thick covering over him, and envelope him with it ; then douche him with water.   Only after the flame is thus extinguished, and the hot clothes have become cool, may they be removed from the body.   If a person has been scalded by hot liquids, pour cold water at once over his body.   If burns are caused by acids (vitriol, *i.e.*, sulphuric acid), or by alkalies, then douche the burnt parts at once freely with water, in order, by greatly diluting these substances, to prevent further injury.

Alleviation of violent pain is the first requirement of the sufferer, and this is done externally by applying cold compresses.   Compresses with spirits of wine will operate as an anodyne, and, when early employed, will even prevent the formation of blisters.   The compress is thus made : cover the burnt part with wadding, pour the spirit over this, and carefully wrap the whole in waterproof material.   As the spirit is very inflammable, in using it the greatest caution is requisite. With slight burns, it is enough to paint them with collodion. If blisters are raised, pierce them several times with a fine sewing needle, but so as not to prick the skin under the blisters.   The effusion of the light-yellow water reduces the tension, and allays the smarting pain.   If, after some hours, the first violent pain is over, it generally does good to besmear the places thickly with almond or olive oil—then cover with a layer of wadding finger-thick, and fasten loosely with a bandage.   Eschar formation always requires medical treat-

* " The hard, rough, black or grey slough produced by caustic or cautery, or left from scalds, or burns," etc.—MAYNE.

ment. In redness, and in blister formations, besmearing
with oil is discontinued about the fourth day—the burnt
place is wiped carefully with wadding, and rice-starch powder
is sprinkled over it; for one or two days more, a thin layer of
wadding is applied until the dried skin falls off.

Burns may be caused *by lightning*. If the person struck
give signs of life, treat him in the same way as one who has
fainted, and the burnt places on the surface of the body as
other burns.

In severe cases of *frost-bite*, the doctor alone can determine
what must be done.

Chilblains are very painful and irritating, especially in bed
at night; on the hands, moreover, they are very disfiguring.
A tepid foot- or hand- bath used every evening and morning,
to which from 1 to 2 tablespoonfuls of nitric acid have been
added, has a soothing effect. In slight cases, after drying the
skin, apply fresh lemon juice; in severe, paint the red places
with tincture of iodine, but not too much; after this is dry,
put on stocking or glove. The painting may be repeated
every second day.

A remedy that may be applied with advantage in all inflammations of the skin, lesser wounds, and sores, and is never
injurious, is *zinc ointment*—this can be had of all chemists,
and should never be absent from any family medicine chest.

Persons *seriously injured*, without external wounds, often
lie in a swooning condition for a time, and must be treated
accordingly (p. 246) until the doctor's further orders. In
lifting and carrying them the nurse, from her experience, will
be able to advise (p. 72).

In *slight contusions*, without wound or injury to bones and
joints, ice-water compresses, wet packings, then dressings with
lead-lotion (a tablespoonful of solution of subacetate of lead,

mixed in a wine-bottleful of water; let none of it get into the
eyes or the mouth) are mostly used. Arnica (used for compresses in some districts) is best avoided, because it sometimes
provokes cutaneous eruptions *(eczema)*, which are worse than
the original injury.

With *wounds*, attention must be chiefly given to the con-
tingency of severe hæmorrhage. Above everything, care must
be taken to prevent the wound being contaminated and
smeared with all kinds of popular impure remedies (cobwebs,
glue, urine, cow-dung). To stop the bleeding the best
remedy is, to close the wound by pressure with a clean finger;
this must be continued until the doctor comes, particularly if
pulsating vessels *(arteries)* bleed violently, when the blood
sometimes spurts from the wound. If it cannot be closed by
the hand, the nurse must know how to compress the chief
arteries (of the extremities) which supply the bleeding part
with blood; this she has to learn in the hospital. If the
bleeding is not too severe, it is enough to apply pieces of clean
ice, or frequently renewed clean compresses dipped in cold
spring water. If it be a slight, superficial skin-wound, after
the bleeding has ceased the parts surrounding it must be
carefully dried, and the wound itself closed with court plaster;
over this, collodion is painted to prevent it from being easily
washed off.

Sometimes, even in rather severe hæmorrhages from wounds
in the hand or in the foot, very high position of the limb is
sufficient. If a long time elapse before the doctor comes, or
should the nurse get weary from keeping the wound closed or
compressed by the hand, and no one is near to whom she
can quickly shew how, properly, to hold the wound so that
the compression (pressure) is continued, then she applies a
bandage (an elastic bandage is best) around the bleeding limb,

and draws it tight until the bleeding stops. This must not be done except when urgently necessary, and then it is better to apply it above the wound (between the wound and the trunk) than upon the wound itself. The nurse now cleanses the wound and the parts around it with clean water ; and if she can procure some carbolic solution (in urgency, even lead-lotion, or a solution of alum may be used), she saturates with it clean wadding or perfectly clean linen, lays it upon the wound, and then bandages the arm or the leg from below upwards to the bandage first put on. But such a bandage must not remain on the limb longer than four hours at most, and it must not be too tightly applied, or mortification may ensue. This proceeding on the part of the nurse can only be justified when a doctor cannot quickly be had, and when the sufferer is already much exhausted—this she will see from the paleness of the face, the coming on of fainting fits, and the smallness of the pulse.

When necessary, the bleeding wound may be subjected to direct elastic pressure—for instance, by a suitable, large bath-sponge being pressed on the wound by a bandage over it ; but this may not be done until the wound has been protected against contact with the sponge and the germs of infection it may contain by means of disinfected dressing gauze, laid direct upon the wound, with waterproof material covering it. Even slight hæmorrhages may become dangerous if long continued, especially with persons described as " bleeders," with whom, from this cause, a small incised wound, or the drawing of a tooth, may endanger life. In such cases, medical assistance cannot too quickly be summoned.

Poisoned wounds. These words generally suggest snake bites, stings of scorpions and the like ; but the stings of *fleas, bugs, flies, and wasps,* are also poisonous—the slight injuries these produce on the skin are so small that they would have no noticeable results were not peculiar, very irritating substances inserted with the bite. When the skin surrounding a fly-sting turns dark blue, and is encircled by small vesicles, medical aid is at once necessary. To bee and wasp-stings, after the stings have been extracted, apply cold compresses and zinc ointment, followed by the same remedies as used with burns ; in severe redness of the skin and swelling, with violent pain, the application of grey (mercurial) ointment quickly relieves and reduces the inflammation.

With us *snake-bites* are rare, because the poisonous snakes are few, and when they do occur they are not so dangerous as bites from the poisonous snakes in hot countries. It is quite a rational, popular remedy at once to suck such wounds (by the uninjured mucous membrane of the mouth and stomach the poison will not be absorbed), and to put a bandage on tightly above the wound ;—but this is useless unless done directly after the bite. The venom can also be drawn out by repeatedly pressing a piece of loaf sugar upon the wound. A little later, it is advisable to wash out the wound with strong vinegar, or strong salt water, and afterwards to bandage it loosely. If liquid ammonia *(solution of ammonia)** be at hand, it may be dropped into the wound.

The bites of dogs or cats suffering from *rabies* should be treated in the same way. It is very important that the person bitten should be set at ease concerning possible results, for, of many hundreds so bitten, only a few suffer from hydrophobia. If the wounds bleed, let them bleed freely—then wash them out and bandage loosely.

With cooks, as well as with butchers and pork-butchers, it

---

* In cases of snake-bite, solution of ammonia is much recommended for external and for internal use, — R. G, often happens (as already mentioned, p. 231) that *juice of slightly putrid meat* penetrates small scratches of the fingers, chiefly at the nails, producing severe inflammations, with swelling of the axillary glands, and high fever. Occasionally this manifests itself directly after the penetration of the juice by violent smarting, and by this, possibly, attention is first drawn to the existence of a small sore place. Immediately washing the wound, and applying compresses with 2 per cent. solution of carbolic, or with strong lead-lotion (2 tablespoonfuls solution of subacetate of lead to half a wine bottleful of water) with complete rest, and a high position of the hand, will often prevent the septic poison from spreading. If, notwithstanding, the redness and swelling increase, rub in grey ointment. But if, from such a cause, the inflammation of a finger continues more than three days, always consult a doctor ; not only unutterable pain, but loss of a finger, nay, even of the hand, of the arm, of life, may result from such poisoning by putrid meat. "Bad fingers" are sometimes caused by fine splinters, which, unnoticed, stick fast in the epidermis and open the way for infectious germs. If a finger pains, and the spot which is painful under pressure be sought with a tiny rod, perhaps a knitting needle, possibly, by carefully removing the (insensible, bloodless) epidermis in layers with a sharp knife, the splinter may be discovered, or an outlet may be made for the drop of pus already formed, and thus worse results be prevented.

HÆMORRHAGE.

Of hæmorrhages that occur without wounds *bleeding at the nose* is the most frequent, and this, principally, among young people and children. During the bleeding of the nose it is not judicious to hold the head bent forward over a wash-hand basin, because, in this position, the natural flow of the blood from the head to the heart is checked. Uniform circulation of the blood is most easily induced when a person lies quietly upon his back with the head a little raised. In so lying, the person bleeding at the nose will get some blood into the throat, and will swallow, or spit it out, but this does not matter. Cold water should be drawn into the nose; to this, some vinegar or alum (a small teaspoonful to a quart of water) may be added. All blowing, squeezing, and wiping of the nose must be strongly forbidden—on the contrary, drawing the breath deeply is to be recommended. As the hæmorrhage usually has its source in the foremost section of the nasal cavity, it is often staunched by simply pressing together the nostrils with the fingers, or by placing a plug of wadding in the bleeding nostril. When, in spite of this, blood still flows backwards into the throat, and when the other remedies specified do not cause the bleeding soon to cease, the doctor must be fetched.

In violent *discharge of blood by coughing, i.e.,* hæmoptysis,* that the blood comes from the lungs or air passages is known by its mostly bright red colour, and from its being ejected with moderate (not always violent) coughing movements. The doctor must be immediately fetched. The patient, above everything, must instantly lie down, and so remain, with the upper part of the body slightly raised; he must keep perfectly quiet, not speak, nor breathe too deeply, and must avoid everything that might provoke coughing. When he is not too weak, lay a large cold compress on his chest, and give him ice to swallow. Then let him swallow from one to two

* Hæmoptysis: "A spitting of blood; hæmorrhage from the lungs."—MAYNE.

teaspoonfuls of salt in a little water ; then give him an injection
of water, to which two spoonfuls of vinegar have been added.
Usually all hæmorrhagic patients very easily become greatly
excited ; for this reason they must be soothed.   Even when
such hæmorrhages from the lungs often recur, as happens in
many cases, they are fatal only in the last stages of consumption.

Vomiting of blood.   In hæmorrhages from the stomach, the
blood ejected from time to time by vomiting is usually very
dark, almost black-brown.   The patient must be kept in bed—
let him swallow small pieces of ice, and apply a cold compress to the gastric region.   If fainting fits occur, no internal
remedies must be administered ; rub the temples with eau-de-
Cologne, give an injection with wine, and apply mustard
plasters to the calves of the legs.

For *intestinal hæmorrhage,* see on Typhoid (p. 204).

Other abdominal hæmorrhages must be treated by the
insertion of pieces of ice, and by injections of ice-water, until
the doctor comes.

ATTEMPTS TO RESUSCITATE IN HANGING AND IN DROWNING.

When there is still some heat in the body of a person who
has been cut down from a halter, or in that of a person
drawn out of the water (a rectum temperature of less than
80.6° F. = 27° C., is certain proof of death), and when, on
laying the ear over the region of the heart, a trace of heart-
beat is still perceptible, attempts to restore animation have
some chance of success.   Besides the low temperature of
the body, the cessation of respiration, and the stopping of the
heart, further sure proofs of death, capable of verification by

the unlearned, are, when vigorous rubbing of the skin, and dropping of hot sealing-wax upon it (on the thigh or abdomen, not upon chest or face, where, should the apparently dead person be resuscitated, disfiguring scars might be produced), produce no redness ; and when, after tying a piece of string around a finger no bluish-red colour appears, and, after removal of the string, the string-furrow does not colour red again but remains white.

A person drawn out of the water must be first laid quickly on his side, and then lifted by the legs, so that the water shall flow out of the mouth and from the respiratory organs.

If respiration and heart-beating be still present in ever so slight a degree, then, by holding ammonia or smelling salts to the nose, besprinkling with cold water, titillating the nose with a feather and the throat with a finger, rubbing the skin, and by other irritants, endeavour to produce movements such as sneezing, vomiting, and coughing, so that respiratory and heart-action may be again excited. If these efforts prove futile, the attempt must be made to produce *regular respiratory action by artificial means.* From my experience, this is best done by blowing air into the lungs regularly, followed by compression of the thorax. A nurse trained in the hospital will probably have seen this operation — it is best accomplished by means of a laryngeal tube, or by a thick canula, introduced into the windpipe ; even without instruments, the blowing-in of air is effectual in many cases. Place your mouth to the mouth of the person apparently lifeless, and, keeping his nose closed, breathe vigorously into him. That the air enters the lungs is proved by the heaving of the chest, which is immediately grasped on both sides by another person, and compressed with a jerk. These operations are carried on alternately, at regular intervals, as in the ordinary breathing of a healthy person. This process must be continued for a quarter of an hour or more, and must be all the more earnestly persevered in when the apparently lifeless person seems to gasp for breath — then wait, only for a moment, to see if the respiratory action continues of itself. If this is the case, the previously named stimulants are again applied until consciousness returns. To this method the objection has been made, that more air gets into the stomach than into the lungs,—but this is not so in all cases; in one case, it is true, I saw that the stomach was greatly inflated by the air blown in, but violent vomiting movements afterwards came on, followed by respiratory and heart action. As lungs, heart, and stomach are regulated by the same cerebral-nerve, it is self-evident that the excitation of a branch of this nerve reacts on its trunk and roots at the nerve-centre in the brain, and, from thence, excites the other branches also. For this reason, in cases of seeming death, even blowing of air into the stomach may be of use—it can in no way injure.

As to other methods of artificial breathing: from the rhythmical (in regular pauses, answering to the frequency of ordinary respiration) turning of the body on to the chest and side (Marshall Hall's method), I have seen no results, but I have from the following method (Sylvester and Pacini's system) of artificially setting the respiration in action : lay the apparently lifeless man on his back upon a table, so that the head almost hangs over one end of it ; then, standing at the head, grasp the arms close below the elbow-joint and draw them upwards until they lie one on each side of the head (first measure, *inspiration*); then carry them again downwards upon the thorax, and with the upper arms and bent elbows of the body compress the lower part of the thorax on both sides (second measure, *expiration*) ; this is repeated continuously, at regular intervals, as in healthy respiratory movement. If properly done, the flowing of the air in and out of the larynx may be distinctly heard. From its simplicity, the nurse can easily employ this method by herself, without help from others, and it is necessary she should practise it.

If *food*, or other *foreign substances* (beans, beads, coins, buttons, plum-stones), should stick fast *in the throat, in the gullet, in the larynx, or in the windpipe*, of children when eating or when playing, then, as a rule, the involuntary retching movements are much assisted by putting the finger into the throat (sometimes the finger reaches the foreign substance and can draw it immediately out). By vomiting, with which fits of coughing easily unite, such substances are ejected, even from air passages. If not immediately removed, fetch the doctor quickly.

If children in play should get foreign substances fixed in the *nose* or in the *ear*, beware of attempting personally to work them out, but immediately send them to the doctor; their sticking fast for an hour longer is of little consequence. Generally, such substances are only pushed in more deeply by ignorant attempts at extraction.

The various methods for rendering assistance in *fits of fainting*, in *apoplectic attacks*, and in *convulsions* will be found in Chapter VII.

POISONING.

It is not always easy to detect whether certain disease-phenomena are produced by the unconscious taking of poisons, or by their secret use—this can only be done with certainty by a doctor. Nevertheless, there are cases in which poison is accidentally taken, and the kind of poison can be directly stated.

In *all cases* in which poisonous substances (mostly liquids) have been swallowed, utilise the time, before the doctor comes and before so-called antidotes can be fetched from the chemist, in getting the poison quickly from the stomach, and, by diluting and intermixing it with oil and glutinous mucilage, in making less nocuous that which possibly remains. To produce *vomiting*, therefore, must be always the first thing aimed at.   This is done by tickling the throat with the finger, or with a feather, and by giving some tepid oil and water mixed *(only in poisoning by verdigris, or by phosphorus, must no oil be used)*.   Then let water, or, better still, milk, be copiously drunk—these cannot injure in any case.

It frequently happens that children, when in the kitchen, take hold of a vessel containing alkali or sulphuric acid * (vitriol), supposing that it contains water, wine, or beer, and drink; although they may desist after the first gulp, even that is sufficient to produce extensive burns in the throat, which may cause death, or at least may be followed by considerable constriction of the œsophagus.

In poisoning by *alkalies*, especially *caustic potash*, administer vinegar; lemon juice; milk; raw eggs.

When strong *acids* have been swallowed, such as sulphuric acid (vitriol), hydrochloric acid, or nitric acid, the remedies are : chalk (prepared chalk), or calcined magnesia mixed in water; soapy water; milk.

In *poisoning by arsenic:* iron-rust or calcined magnesia mixed in water; milk; raw eggs.

In poisoning *by verdigris:* much sugar and eggs; *no vinegar, nor oil, nor fat.*

---

* Solutions of caustic potash are used by laundresses, and sulphuric acid is often used for cleaning metals.—R. G.

In poisoning *by phosphorus : no oil ;* much magnesia and water.

In poisoning *by opium and morphia :* much strong black coffee ; sour, red wine.

In poisoning *by prussic acid* (in bitter almonds, cherry, apricot, and peach stones), and in poisoning by *cyanide of potassium :* black coffee ; cold douching of the head.

In poisoning *by berries of the deadly night-shade (Bella-donna)* : emetics ; cold douching of the head ; lemon juice and water ; coffee.

In poisoning (suffocation) *by coal gas, or carbonic acid gas :* first of all, remove the sufferer from the room into which the gas had escaped ; then, to recover animation, take the same steps as in the case of one seemingly dead : besprinkling, douching with water, rubbing, injections with vinegar, mustard poultices. Long-continued artificial breathing has proved, in these cases, particularly efficacious.

CHAPTER IX.

FOOD AND DIET.—CONCLUSION.

THE excellent books of Dr J. Wiel ("Diätetisches Koch-buch," and "Tisch für Magenkranke"), which should be in every hospital and in every family, relieve me from the task of here entering into particulars of the diet fitted for distinct diseases, or the method of preparing particular foods. A housewife, at home in kitchen and cellar, will readily understand the practical instructions given in these books; but a "fine" lady, and a nurse unaccustomed to cooking, cannot acquire the culinary art from books alone: they must have practical instruction. Nevertheless, I will here state some of the fundamental principles which govern the selection of suitable nourishment (Dietetics).

Nutrition is a highly complicated process in all living things (organisms: plants, animals, man), *and is dependent upon very many circumstances.* To grow and to sustain itself, every living organism must take substances into itself from without,—from these substances the component parts of the body are formed, and the body increases by their growth. Man absorbs air by his lungs in breathing; the constituent parts of plants and animals, as well as minerals (mostly in the form of mineral salts), he absorbs by the stomach and intestine. In order that a certain part of the food eaten shall become blood, all food constituents must be brought into a liquid state, if not liquid when taken. From the blood flowing through them, the soft and hard parts of the body absorb what is necessary for their support, and, in addition, in the young for their growth.

Properly to effect this the following is necessary: the food, only imperfectly reduced by being cut into small pieces, must be changed, by *mastication*, into a pulp-like condition, and be *mixed with saliva* before entering the stomach. Even in this stage it may be a source of bad digestion and bad nutrition. Toothless, particularly aged people, suffer from digestive disturbance because they swallow the food in too large unmasticated pieces. Children are often too indolent to masticate sufficiently, or they are too animated, too diverted to devote the requisite attention to it. For some weak patients, masticating is just as fatiguing as if they were required quickly to lift their arms high fifty times in succession. Toothless people are helped by artificial teeth, by cutting the meat into small pieces, or by mincing it; when necessary for sick patients to take solid food it must be minced. Children are often troublesome when eating, but this must be corrected in their training. To give patients nothing but minced meat is not desirable, it soon becomes repugnant to them, and the secretion of saliva, so important to digestion, is too little stimulated. Saliva is absolutely necessary for digesting bread, farinaceous foods, and potatoes. With persons of weak or of indolent mastication, the bread should be given dry, in the form of rusk or of biscuit; these crumble quickly in the mouth, or they may be dipped in milk, tea, or bouillon, and thus eaten; in this form bread easily dissolves in the stomach, but much new, soft bread, consolidates into hard lumps, upon which the digestive juices operate slowly and with difficulty. With children, occasionally, *the swallowing of masticated food* requires great effort, at times is almost impossible, especially when the food has been pushed into the sides of the cheeks and has formed hard lumps. This bad habit of the child should be corrected during his eating.

After its entrance into the stomach the food is at once *operated upon by the gastric juice*, and, subsequently, by *the intestinal juice*. These juices proceed from the mucus of the inner coats of the stomach and intestine *(mucous membrane)*, and reduce a great part of the (digestible) foods to a gelatinous, and later, to a liquid, state, in which form they enter the intestine, are absorbed *by the lymphatic ducts of the intestinal mucous membrane, and, through them, are conducted into the blood; in their progress they pass through extremely fine sieves—the mesenteric glands.* The derangements that may occur in this process are very manifold.

1. *Too little*, or *too much*, *gastric juice*, *or a gastric juice too weak in its action*, *may be secreted*, which may arise from very different causes. Above everything, the formation of healthy gastric juice demands a copious supply of healthy blood to the stomach, and unhindered active motion in its blood-vessels; for this purpose, *healthy gastric nerves are essential.* Anæmic persons, feverish patients, or those who suffer from exhaustion, or from nervous causes, secrete an insufficient quantity of gastric juice; occasionally, in typhoid patients, the secretion of saliva and gastric juice is wholly suspended. The mucous membrane of the stomach may itself be diseased, and therefore be inactive. In most books on dietetics too little attention is paid to the *influence of the nerves* in this process. It is an old saying, on smelling a fragrant scent of food, "it makes the mouth water." This is almost literally true. Fragrant foods, indeed even the vivid idea of such, rapidly produce increased secretion of saliva, as well as of the gastric juice. How much the state of the mind influences the appetite is well known. Then again

the instinctive impulse of imitation in man when eating in company must be taken into consideration—every house-wife has doubtless observed, that some indifferent eaters influence prejudicially the appetite of the rest of the guests ; whilst on the contrary, children who, ordinarily, can be induced to eat with difficulty, will take double and treble the quantity when their ambition is incited by seeing other children, of like age, eating well beside them. How much the appetite is stimulated by the pretty appetising arrangements of the dishes, of the dining-table, of the dining-room, with its comfortable seats, &c. !

Although the doctor must always be consulted as to the degree to which the stomach of a patient, suffering from loss of appetite, may be stimulated by so-called piquant foods, yet their use in moderation is permissible. To some patients the scent of food is particularly offensive, and especially that of hot meat ; they would rather have cold roast beef and ham than the finest and most tender hot roast meat. Habit, also, and personal inclination must be yielded to, so long as they are beneficial. He, to whom roast meat was repugnant in health, must not be compelled to partake of it when ill.

2. *The gastric juice must be able to come into close contact with the food.* This is impossible when the stomach is over-loaded, either with large quantities of fluids, or with masses of chyme ; and further, it is not possible when the diseased stomach secretes much mucus. For this contact it is necessary that the stomach should contract itself, should mix the foods, and should propel a portion forward in order to be able to operate upon the rest. Flaccid, almost paralyzed, coats of the stomach, may thus (for instance, in dilatation of the stomach) cause bad digestion and bad nutrition.

.3. *Food taken in large quantities, remaining in the stomach and there fermenting,* exercises a morbific influence upon its mucous membrane, and renders the gastric juice inoperative.

4. *The quantity of the dissolved nutriment absorbed depends upon the activity and number of the lymphatic ducts in the intestinal villi.* This activity again is influenced by like circumstances to those which influence the activity of the mucous membrane of the stomach.

5. *If the mesenteric glands,* through which the chyle must pass from the lymphatic ducts, *are obstructed* (as frequently happens with children), starvation ensues, because very little chyle, or none at all, gets into the blood.

6. *The supply of blood, with which nourishing chyle is combined,* would not have the slightest influence upon the other parts of the body unless such parts were able to absorb that which was supplied, and to produce therefrom that which they themselves are : *muscle* (flesh), *adipose tissue, connective tissue* (membranes), *glands, brain, nerves, vessels.* The essence of life, therefore, lies as much in these tissues as in the cerebral and respiratory systems. Not one of these functions can exist without the other ; not only the whole man, but the very smallest part of him lives. For instance, if a contrivance could be made by which, from the heart, the blood and the chyle could be put in motion in a dead body, the dead body would not live, because the separate particles (organs, &c.) had previously lost their vitality.

How very intricate, and how very difficult all this is to understand! The reader should not worry himself with the desire to comprehend it fully. I have given him only a

* " Small conical projections . . . having small pores which are the mouths of the absorbent vessels."—MAYNE.

glimpse into this small part of the science of life (biology, physiology), that he may not take it to be a simple matter to make a lean person fat, or a weak strong, or a fat thin. It is the doctor's duty to study each case, to find out wherein the derangements of the nutrition lie, and the best means to cure them.

In conclusion, a few general pertinent remarks must be made.

It is evident, from what is here said, and previously on feverish diseases, that *certain natural limits are set to the alimentation of the sick.* The idea of desiring to remedy, even by compulsion, all emaciation and weakness by giving large quantities of agreeable food, must be abandoned. Apart from the consideration that, in many cases, it would operate with positive injury and even fatally, the taking of large quantities of food by a person diseased in the organs of nutrition will avail just as little as if the most sumptuous dinner had been poured into an automaton.

Fortunately, complete failure of the function of digestion is not frequent ; even in the most serious diseases, the taking of that which is the most necessary of all to life, water, or of thin solutions of some important kinds of food, is still possible. It is amazing to the uninitiated to learn how long the lives of men may be thus protracted, who (often in typhoid) lie unconscious for weeks insensible to hunger and thirst, and who, without care, would perish.

The natural philosopher well knows that a temperate man, living in good circumstances, is wont to eat at least from four to six times as much food as is necessary for the due discharge of his daily duty. Man is distinguished from the animal in that he not only eats to live, but because eating is a source of enjoyment to him.

It is a wide-spread error to suppose that a *very concentrated liquid food* will quickly give strength to delicate children and weak patients. This is usually frustrated by the resistance of the sick, who refuse with aversion the strongest and most savoury broths—to whom, even soup, as it usually comes at other times to their table, is too strong. Excepting solutions of different mineral salts already contained naturally in most aliments, there are no foods that can be absorbed into the blood without joint action of the saliva, gastric and intestinal juices, and without movement of the stomach and intestinal canal ; if the digestive material become exhausted, and if the movements of the digestive organs be suspended, then the introduction of food can only encumber the viscera. Between this complete inaction and the vigorous healthy activity of the organs of nutrition, there are as many intermediate stages as there are between positive disease and perfect health.

Man reluctantly descends the many little stages of reduced health, such as lowered strength, and lessened activity, because he instinctively feels that, after descent, he seldom regains the level at which he stood before. If circumstances thrust him down several stages at once, or if, in wantonness, he himself leaps down, then he cannot rise again with a bound, but he must get up step by step, slowly, as a weak helpless child ! The feeble patient must return almost to the food, and to the habits of a child ; food will only be good for him when given frequently, never too much at once, and in a diluted form adapted to his weak digestive powers. If, every hour or two, some pigeon, chicken, or veal broth is given to a convalescent, then some tea, with rusk and such like, all having very trifling nutritive value, by this means he will not only become stronger daily, but he will look forward to his little meals with pleasure. If he has the most concentrated beef-tea, ox-tail or turtle soup once

or twice a day,—these, for a healthy person, are like liquid
meat, and to them also the highest nutritive value possible to
human food in liquid form can be given by all sorts of additions,—then the patient, with reason, will resist taking even a
spoonful of it (its strong aroma, delicious to a healthy person,
is most objectionable to him) ; if he took it, it would lie in
his stomach like lead, without doing him any good.   In
feeding and dieting patients we must remember, that civilized
man has not chosen his mode of living because he considers
it the most healthy (for instance, he takes his meals two
or three times a day, and then in considerable quantities),
but because it enables him, under these conditions, most
surely to gain the necessary money to live, and the most
time for his work.   Just look around !   Where are the people
who ask, which country, which town, which occupation, what
manner of life will be most conducive to their health ?   How
small is the number for whom these questions could come
under consideration at all !   Most people live as, according to
their circumstances, they must live.   If the body suffers, and
is worn out after years of work, and is no longer as capable
as formerly, then the help of the doctor is sought.   If
medicine be prescribed, no matter how expensive, it is pro-
cured and regularly taken.   But when the doctor explains to
the applicant that he must alter his way of life in certain
directions, and makes him understand that he is only partially
able to work at all, and this partial ability will only continue
so long as he does this and avoids that, then the patient
will scarcely listen to the doctor, and will either declare
simply that he cannot, and will not, change anything in his
way of living ; or, that, in his opinion, it is an error to suppose
that he is really ill : if the doctor would but give him an
efficacious remedy for his head-ache, and for the pain in his
stomach, he would be quite well again.

That epidemics and some other diseases strike men from
without, suddenly like an accident, is well known. That,
in many more diseases, the germ is inborn in the weak
development and imperfect maturing of this or that part
of the body, and, that complete uniform health of the whole
body and mind is just as rare as perfect beauty in every
part, many admit as true indeed in general, but they are
not willing to recognise that they themselves belong to
the imperfect creatures, and will so belong for life. But
this should be required of no one ! The idea that all men
are equal is one of the most beneficent delusions by which
the Christian world, in this century especially, is permanently
ennobled and perfected,—it is the source of immeasurable
happiness to many thousands, because to its acceptance is
closely attached the claim of equal rights for all men, which
forms the basis of our present human society. I will not
disturb this idea; but it is well to think clearly upon the
subject, because, then only will the proper means be found
by which to balance the inequalities again and again produced
by nature. That these inequalities are often made manifest
for the first time when the powers are tested by unusual work,
and that even the strongest man, by excessive effort, or by
age, sooner or later finds his capacity for labour reduced, is
a sorrowful experience, to which man must accustom himself
with resignation.

It is certainly a misfortune to be able only to live as a sick
man, or as a feeble valetudinarian, imperfectly fitted for work
or for enjoyment. But it is equally painful to both doctor and
nurse when they are able to render but little assistance, or
none at all. In this sorrowful dilemma they have one con-
solation : the knowledge of *having conscientiously fulfilled their
duty to the fullest extent of their power.*

APPENDIX.

THE STRUCTURE AND FUNCTIONS OF THE HUMAN BODY.

THE following is designed for the Instruction of Female
Pupils in the Preparatory Course on the Care of the
Sick, who, by lectures, have already learnt something of the
subject, and have also seen anatomical preparations of, at
least, some parts of the human body now to be considered.
The particulars, difficult at first of comprehension, must be
strongly and clearly impressed upon the memory by repeated
reading. To render the probationer's examination for her
certificate less difficult, a great part of what follows is given in
the form of question and answer. The "Anatomische Wand-
tafeln für den Anschauungsunterricht" (Anatomical Wall-Plates
for Objective Instruction) of Dr Hanns Kundrat, published
in Vienna, price six florins, may be recommended as suitable
illustrations.

GENERAL INTRODUCTORY REMARKS.

Anatomy is the science of the structure of the human
body.

Every part of the body which performs a certain action
(function) towards the maintenance of the whole is termed

an *organ.* The science of the actions *(functions)* of the organs of the body is termed " Physiology."

For many functions the joint action of several organs is necessary : a *system of organs.*—Example : bones, muscles (flesh), and motor nerves acting together bring about the movements of the body—they are *organs of motion ;* conjointly, they are termed *the motor system.*

THE INTEGUMENTS.

All integuments are composed of very fine filaments (membranous tissues).   There are different kinds of membranes, but, of these, two only will here be considered :

1. *The skin, or cutis* *—This completely covers the body ; in it the hairs, to it the nails, are fixed.   The skin secretes sebaceous † matter, which covers the surface as a thin layer, and is slowly but continuously renewed ; in addition, on the body becoming heated, the skin secretes perspiration. The skin varies much in thickness ; is thickest on the back, thinnest on the eyelids.   Directly under the skin, and intimately united with it, is the fat, also termed *adipose tissue,* because the fat lies in fine honeycomb-like cells, which again consist of thin membranes and filaments.

2. *The mucous membranes.*—This term denotes such membranes as secrete mucus ; they are always much softer, looser, and redder on the surface than the epidermis, and have no adipose tissue under them.   The oral cavity and the tongue are covered with mucous membrane, as is also

---

* "The skin, as commonly regarded, is composed of three membranes : the outermost is the scarf-skin, cuticle, or *epidermis ;* the middle, the *rete mucosum ;* and the innermost, the true skin, *cutis vera* or *derma.*"— MAYNE.

† "Sebaceous : fatty, suety."—MAYNE.

the interior of the windpipe, of the œsophagus, of the stomach, of the urinary bladder, &c. At the lips, at the edge of the nostrils, at the anus, at the orifice of the urethra, the mucous membrane changes into cutis.

THE SECRETORY GLANDS.

By *Secretory Glands* is understood the organs whose function it is to prepare (to secrete), from the blood flowing through them, a special juice peculiar to each gland. These glands have *excretory ducts*, conducting to, and discharging at, the parts to which the juices flow to their designed use. The *largest gland* in the body is *the liver;* it produces bile ; the bile next collects in a bladder *(gall bladder)*, and flows thence through an excretory duct into the intestine, where it is necessary to digestion. The *salivary glands* (two in the face below the ears, two under the tongue, and two in the throat under the jaw, all discharge into the mouth; and one pancreas, which lies behind the stomach, and discharges into the intestine by the side of the gall-duct) produce the saliva, likewise necessary to digestion. The *kidneys* are *urinary glands;* they produce the urine, which is conducted by their excretory ducts *(ureters)* into the urinary bladder. The *sebaceous matter* of the skin is secreted by small glands in the *cutis vera*, in size scarcely as large as grains of millet; and the *mucus*, by glands in the mucous membrane as large as peas, and so on.

THE BLOOD AND THE LYMPH. VASCULAR AND LYMPHATIC GLANDS.

The red, easily-coagulable fluid, universally known as blood, flows through the body in membranous tubes, of which the largest are thumb-thick, and the smallest finer than the finest hair.

The *spleen* (lying in the upper left side of the abdomen), and the *thyroid gland* (lying in front of the larynx, and which, when enlarged, is termed *goître*), are both immensely sanguineous organs, the functions of which are unknown ; they are supposed to have some relation to the formation of blood, and are termed *vascular glands*.

By *lymph* is understood the colourless fluid which partly flows slowly through very fine tubes, and partly fills the finest network of the organs and tissues of the body. It collects, in many parts of the body, in small spongiform corpuscles termed the *lymphatic glands*, particularly at the throat, in the armpits, the groins, and in the mesentery. In health, these glands are soft, scarcely as large as a pea, not easily felt externally ; in disease, however, they may swell to large protuberances. A part of the food eaten is not absorbed from the intestine directly into the blood-tubes *(blood-vessels)*, but into the very many lymph-tubes *(lymphatic vessels)* — arranged net-like — of the intestinal mucous membrane ; from them, the lymph (here termed *chyle*) flows into the lymphatic glands *(chyle glands of the mesentery)*, and from these is formed, like a tree from its roots, a fine membranous tube, which terminates in a vein at the throat, and through this tube the lymph is conducted from the intestine into the blood.

N.B.—After the pupil has read the sections on the Alimentary Canal and the Position of the Viscera, this chapter must be re-read.

THE MOTOR SYSTEM.

A.—THE BONES.

1. *What are the different kinds of bones in the human body?*

There are *long* bones, also termed *tubular* bones; these are located in the limbs (extremities), *e.g.*, one in the upper arm, one in the thigh, two in the fore arm, and two in the leg;—*laminated* bones, *e.g.*, bones of the head (skull), the pelvis, the shoulder-blade; and *short* bones, *e.g.*, the vertebræ, the wrist, and the ankle bones.

2. *Are the separate bones equally solid in all their parts?*

No. Most bones, specially the tubular bones, have a hard outside shell, and are filled with adipose tissue (bone-marrow). The short bones are formed internally like a hard sponge; bone-marrow lies in their small interstices.

3. *How are the tubular bones joined?*

Their ends are thick, often spherical in form, and also concave (pits, sockets). The spherical projections of one bone fit into the concavities of another, and are united to them by firm fibrous ligaments. This kind of bone-connection is termed *joint*, and the connecting ligaments are the *articular ligaments*.

4. *What is the specific purpose of the bones of the body?*

They answer different purposes: (1) They form the strong framework *(skeleton)* of the body; (2) The muscles *(the flesh)* are fastened to them, and by the muscles the bones are moved at their joints; and (3) They enclose important organs necessary to life, and preserve them from injury.

5. *Which are the principal parts of the bone framework,* (skeleton) *and of the human body as a whole?*

The head, the trunk, and the limbs or extremities.

6. *Of what does the bony part of the head—the skull—consist?*

The *skull* is a bony sphere, in which the brain is located; it is composed of several lamellar curved bones joined (*sutured*) firmly together by their finely serrated edges. To the lower and front parts of the skull the bones of the face are firmly and immovably united, with the exception of the lower jaw—this is fastened to the skull by a movable joint (*temporo-maxillary joint*) situated in front of the ear. At the base of the skull is a large hole through which the spinal cord, coming from the brain, passes into the long canal running through the interior of the vertebral column; it has also many small apertures for the exit and entrance of nerves and blood-vessels.

7. *Is the skull of a new-born child constituted as that of an adult?*

No. At the top it is not at first quite ossified,—there, in two places, only a membrane lies over the brain; these soft parts of the skull are termed the *anterior and the posterior fontanelle;* in the course of the first year of life these close, that is to say, these membranous parts of the skull become ossified.

8. *How many teeth form the complete set of an adult?*

Thirty-two; sixteen in the upper jaw, and sixteen in the lower.

9. *Are all teeth alike in form?*

No. In front, above and below, there are four narrow wedge-shaped teeth, the *incisor* teeth; then follows on each side, above and below, a sharp angular tooth named *eye* or *canine tooth;* and these are followed on each side, above and below, by five *molar or grinder teeth*, with broad masticating surfaces.

10. *How are the teeth fixed in the jaws?*

By one, or by several continuations, the so-called roots, which fit into suitable cavities in the jaws. Every incisor, and every canine tooth, has each one root; molar teeth have each from two to four roots. The part of the tooth exposed in the mouth is termed the "crown" of the tooth.

11. *Is the child, at birth, furnished with teeth?*

No. The teeth begin to grow out of the jaw towards the end of the first year of life; first, the incisor teeth, then the front molar teeth, then the canine teeth. These, mostly small and delicate *(milk-teeth)*, subsequently drop out (generally between the seventh and eighth years of age), and are replaced by new. This changing of the teeth (also termed "shedding of the teeth") affects only the incisors, the canine teeth, and the two front molars; the three back molars come once only, and are not renewed. The hindermost molar tooth (the *wisdom* tooth) sometimes does not appear until towards the twentieth year of age.

12. *Of what does the trunk of the body consist?*

Of the vertebral column, the thorax, and the pelvis.

The *vertebral column* of man is composed of seven cervical, twelve thoracic, and five lumbar vertebræ (altogether twenty-four vertebræ). The cervical vertebræ are the shortest, the lumbar the longest. The vertebræ are held together by strong fibres *(ligaments)*; between each two vertebral bodies (by this term is meant, the thicker part of the bone in front of the ring enclosing the spinal cord) a soft cartilage is intercalated. The whole vertebral column is like a flexible staff made up of broad signet rings. In the canal of the vertebral column, which is produced by all the vertebral rings lying one over the other and closed at the back and sides by membranes, is *the spinal cord;* this descends from the brain through

u

the lower cranial aperture, and sends off, in every direction, nerve-strands to the other parts of the body through openings in the walls of this canal.

13. *In what way are the head, the thorax, and the pelvis united to the vertebral column?*

The skull is so united to the upper cervical vertebræ by joints and ligaments that it can be turned, in all directions, up to a certain point.

On both sides, into the lateral parts of each of the twelve thoracic vertebræ, a rib is set. The *ribs* (twenty-four in all) are curved bones of various lengths, which change into cartilage at the front of the thorax, and there unite with a flat upright bone, the *sternum* or "breast-bone," which is about two fingers broad and a span long. The ribs are united to each other by strong fibrous membranes, and form a basket-like case around the lungs and heart, encompassing and covering them. The *thorax* is, therefore, formed by the thoracic vertebræ, the ribs, and the sternum or breast-bone.

The *pelvis* is an oblique-oval ring, formed of flat, broad bones, and narrows towards the lower part; the middle part of the pelvis at the back (the *sacrum*) is fitted to the lowest lumbar vertebra, and is united thereto by strong ligaments.

14. *How are the limbs fastened to the trunk?*

Each thigh-bone, at its upper end, has a spherical head, which head fits into a deep concavity (*socket*) in the pelvis and thus forms the *hip-joint.*

In each arm, the spherical head at the upper end of the bone of the upper arm fits into the shallow socket of the shoulder-blade and thus forms the *shoulder-joint.* The shoulder-blade is a flat bone fastened by muscular tissue to the back of the thorax, at the top of the back; in front, it is united to the *collar-bone,* which lies like a girder between the shoulder and the breast bone.

15. *How are the other parts of the limbs constituted?*

The fore-arms and the legs have each two bones. The bones of the fore-arm are named the *ulna* and the *radius;* at their upper ends *(elbow-joint)* they are joined to the lower end of the *humerus*, and at their lower ends *(wrist-joint)* to the hand. The hand consists of the short carpal bones,[1] the metacarpal[2] bones, and the digital bones[3] (small tubular bones), and these are all connected by joints.

The bones of the leg are named the *tibia*[4] and the (much thinner) *fibula*[5]; they are connected at their upper ends (knee-joint) with the lower end of the *femur*[6]; a bone lies in front of the knee-joint called the *patella*[7]—this is inserted into the lower end of the tendon of the long muscle by which the leg is extended. At their lower ends, the bones of the leg are joined to the foot by the ankle-joint. The foot consists of the short tarsal bones,[8] the metatarsal ditto,[9] and the phalanges[10] of the toes (small tubular bones), and these are all connected with each other by joints.

B. THE MUSCLES AND THE MOTOR NERVES.

1. *What causes the movements of the joints of the bone framework?*

The *muscles.* These are flat or round, long or short, strips of red flesh, firmly attached at both ends to different bones. All that is termed flesh in ordinary life is muscle. Some muscles, specially those at the extremities of the limbs which serve to move the hands and the feet, are affixed to a bone

[1] Wrist bones.
[2] Middle hand-bones.
[3] Finger bones.
[4] Shin bone.
[5] Sural bone.
[6] Thigh bone.
[7] Knee-cap.
[8] Ankle bones.
[9] Instep.
[10] Toe bones.
—Tr.

by their upper ends only; at their lower ends they change into white, shining ligament-like strips, the *tendons*, and it is these tendons which are then fixed firmly to the bones.

2. *In what way does the muscle move itself, and the bones?*

It contracts, becomes thicker, firmer, and shorter in so doing; this action takes place with such force that both the points at which the muscle (or tendon) is fastened to the bone are drawn closer together.

3. *But what is it that causes the muscle to contract?*

A power which, through the nerves, flows to the muscle from the brain and spinal cord.

4. *What is understood by nerves?*

The nerves are milk-white threads issuing from the brain and spinal cord — at first, lying together in rather thick strands, afterwards separating and becoming thinner and thinner, until, finally, they spread (when they are nerves of motion) like a fine network of infinitely fine fibrils, in the muscular tissue; by these nerves the motor-nerve-power is conveyed into the muscles. Without nerves a muscle cannot contract, even when otherwise living and healthy. If the motor-nerve of a muscle be divided by cutting, the muscle is paralyzed.

5. *Where does the motor-nerve-power originate?*

In the brain principally, but also in the spinal cord. When the brain or the spinal cord becomes diseased, and certain parts of the one or of the other are destroyed, the development of motor-nerve-power ceases, and paralysis sets in. But such irregularities may occur in the flowing of the motor-nerve-power that it passes too energetically and continuously into the muscle, and thus "cramp" is produced.

6. *Does this motor-nerve-power always act with the same energy?*

The nerve-power acts upon the heart and peristaltic * movements of the intestines continuously and regularly, without our help. The respiratory movements also continue without our noticing them, consequently during sleep ; but, when awake, to a certain extent we can retard, accelerate, lessen or intensify respiratory action. All other muscles of the body are always in exercise to a moderate degree, though we are unconscious of doing anything to promote their action. But, by the will, we can immediately send more or less nerve-power into the muscles, and thereby, at pleasure, can bring them into slight or vigorous, slow or quick, contraction.

ON THE NERVES OF FEELING AND OF SENSATION.

(THE NERVOUS SYSTEM.)

1. *Have the brain and spinal cord other duties to discharge besides those which produce movement?*

Yes ; in them the sense of feeling is produced by the nerves of feeling, and the sensorial perceptions by the sensory nerves. The brain is also the organ of the intellectual functions ; the brain, spinal cord, and nerves combined form *the nervous system.*

2. *How do the nerves answer sensation?*

The nerves of feeling conduct the stimulus of contact and the variations of temperature to the brain.

Increased, intensified contact, as also increased cold or heat, produce pain. Should the nerves of feeling, which radiate from a given part of the body, be severed, then, in that part, all feeling ceases, paralysis of sensation, *i.e.,* insensibility,

---

* "Applied to the peculiar movement of the intestines, like that of a worm in its progress."—MAYNE.

follows.   Diseases of the nerves of feeling, of the brain, or of the spinal cord may increase the capacity to feel (hypersensitiveness), or may even destroy it (insensibility).

3. *Does the motor-nerve-power, conducted from the brain to the muscles, run in the same channels as the currents of sensation conducted to the brain from without?*

No; there are special motor nerves and special sensory nerves.   But both kinds of nerves often lie (*e.g.*, in the limbs) side by side in one nerve-trunk or bundle of nerves.

4. *How many senses have we? and what are their functions?*

Five senses, as follow :—

(1.) The sense of touch and feeling, already described ;

(2.)   „   seeing ;

(3.)   „   hearing ;

(4.)   „   tasting ;

(5.)   „   smelling.

The interior of the eye, of the ear, and of the nose, and the surface of the tongue are each provided with special kinds of nerves, which nerves have each only one kind of faculty for perceiving and conducting.

By the optic nerves, only light and colour are perceived ; by the auditory nerves, only noises and tones; by the gustatory nerves, only liquid, flavoured substances; by the olfactory nerves, only gaseous (volatile) substances, perceivable by smell.

The destruction of these nerves at their terminations in the organs of sense (eye, ear, tongue, nose), or in their course, or in the interior of the brain, results in optic paralysis (blindness), auditory paralysis (deafness), gustatory and olfactory paralysis (loss of ability to taste and smell).   Also from diseases of the organs of sense their functional activity can be injured.

5. How can all the functions of the nervous system be briefly summarized?

The brain and spinal cord are the parts of the body in which movement originates, and to which sensations and other perceptions are conveyed. In them a certain quantity of motor-nerve-power is always stored, which, partly without conscious incitation, flows on the tracks of the motor nerves to the heart, to the viscera, and to the respiratory muscles, and, partly by the will, is sent to the one or to the other muscles of the body. The brain and spinal cord are also the seat of the perceptive faculty; the influences operating upon the body from without, such as contact, heat and cold, are conveyed to the brain and the spinal cord by the nerves of feeling.

Further, light, sound, specific qualities of drinks (flavours) and of volatile substances (scents) are conducted by special nerves for seeing, hearing, tasting, and smelling to the brain alone, and come to its consciousness equally with the sensations.

Without the living, healthy action of the nervous system, man would be able neither to move nor to perceive anything external to him. By the nervous system alone he comes to the consciousness of a world outside himself, and, by this means, to "self-consciousness," that is, to the perception of his own personality.

THE HEART AND THE BLOOD-VESSELS. CIRCULATION OF THE BLOOD.

(THE VASCULAR SYSTEM.)

1. How is the human heart constructed?

The heart is broad at the top, and almost round; narrower below, terminating in a rounded point; it is composed of muscular tissue, and is divided by a wall *(septum cordis)* longitudinally into two halves—a right and a left; each half is subdivided by a transverse partition wall into a smaller upper part *(right and left auricles)*, and a larger lower part *(right and left ventricles)*. These transverse partition walls have large openings which, during the contractions of the heart, are closed by valves stretched, sail-like, over them *(tricuspid and mitral valves)*.

2. *What are the blood-vessels?*

They are tubes in which the blood runs, and they are formed of membranes more or less thick.

3. *How is the heart connected with the blood-vessels?*

From each ventricle a tube issues, about the thickness of the thumb; the tube from the right ventricle *(the great pulmonary artery)* turns directly across towards the left, divides into two branches, one passing to the right lung, the other to the left. The tube from the left ventricle *(the great corporal artery, the aorta)* first proceeds upwards, and to the right, forms an arch directly over the heart,—and, from this arch, the main arteries of the head and the arms branch off,—then turns towards the back, and runs downwards in front of the vertebral column. In its course it supplies branches to the thorax and the viscera, and, at the base of the vertebral column, separates into two principal branches for the legs.

4. *What is the further action of the arteries?*

The pulsating vessels *(arteries)* divide like the branches of a tree (like the nerves) into ever finer ramifications—enter into all parts of the body, and become thinner and still thinner tubes until the finest *(capillary vessels)* can only be perceived by the aid of a magnifying glass. The blood-vessels have no terminations (as possessed by the finest nerve ramifications), but the finest branches unite finally into net form; from these nets other new tubes form afresh, which then again join others of the same kind, and thus unite together to form ever thicker tubes, until, by degrees, they become just as large as the arteries. These blood tubes, which, to a certain extent, grow out of the capillary network, are termed *veins*; their walls are thinner than those of the arteries, and they are without pulse beat. The veins, which have become large, finally run by the side of the great arteries back to the heart, into which they discharge—that is to say, the four veins, which come from the lungs, discharge into the left auricle; the two veins, which come from the other parts of the body, into the right auricle.

In *consequence of the contractions of the heart* the blood is driven through the arteries, not only until it reaches the finest capillary nets, but through these, along the veins also, and back again into the heart. This movement of the blood is termed *the circulation.*

5. *What is meant by the "less"* (pulmonic), *and the "greater"* (systemic), *circulation of the blood?*

That which is termed the *less circulation* is the movement of the blood from the right ventricle into the great pulmonary artery, then through the capillary vessels of the lungs into the veins of the lungs, and, through them, into the left auricle.—Thence the blood runs into the left ventricle, and is driven through the great corporal artery into the other parts of the body, and, from above and below, returns through two large veins back into the right auricle—this is the *greater circulation.* From the right auricle the blood runs into the right ventricle, and thence again through the pulmonary artery into the lungs, and so the process is continuously repeated.

6. *Does the heart contract in all its parts simultaneously?*

No. Both ventricles contract at the same time, and at the

same time impel the blood into the lungs and into the body as described. Then the auricles contract and drive the blood into the relaxed ventricles. During the contraction of the ventricles the heart heaves somewhat, so that the apex of the heart beats against the front thoracic wall *(impulse of the heart's apex)*.

7. *When the ventricles contract, why is it that the blood does not flow back into the auricles?*

Because, during the contraction of the ventricles, this is prevented by the openings of the auricles being closed by valves. Should this somewhat complicated arrangement be more or less destroyed by diseases of the heart, then irregularities arise in the circulation of the blood, which may become of serious consequence, specially to the respiration.

8. *How frequently does the heart contract?*

From sixty to eighty times per minute. These contractions are not only felt at the apex of the heart, but also by the blood-wave, which is transmitted from the heart into the arteries with the greatest velocity. To ascertain the velocity of the heart-beats (of the pulse), one usually feels the radial artery on the inner side of the arm, close above the wrist on the thumb side.

THE TRACHEA, THE LARYNX, AND THE LUNGS. THE RESPIRATION.

(THE RESPIRATORY SYSTEM.)

1. *What is understood by the Trachea and the Larynx?*

The *Trachea* (wind-pipe) is a tube issuing out of the thorax from the lungs, and, passing upwards inside in the front of the throat, terminates in the mouth close behind the tongue.

This tube consists of cartilaginous rings, united one to another by a fibrous membrane, and, at the top, widens for a length of about two inches; this part consists wholly of cartilage, and is called the *larynx* (head of the windpipe).

Air is drawn into the lungs through the trachea and the larynx, and flows out again through the same channel. A mucous membrane lines both windpipe and larynx throughout. This mucous membrane forms two folds, one on each side in the larynx, called *the vocal chords*. The air issuing from the lungs causes these vocal chords to tremble, " vibrate," and tone—*the voice*—is produced in the larynx. By means of small muscles we are able to expand or relax the vocal chords at will, and, by vigorous expiration, can produce different kinds of noises and tones in screaming, singing, or speaking. By giving different positions to the tongue, the palate, and the lips, the manifold forms of tone and sound of which speech is composed are produced.

2. *What is the nature of the Lungs, and how are they connected with the Trachea?*

Within the thoracic cavity man has one lung on each side. The lungs are composed of fine membranes and fibres ; in their structure they may be compared to a sponge with very fine pores, encased in a fine membrane *(pleura-pulmonalis, or pleura)*, which sponge, one must imagine, is capable of expansion so that it can be inflated.

The trachea is first divided, at its base, into two main trunks, one for each lung ; each trunk sinks in into its own lung, where it divides itself again and again into thinner branches, until these finest tracheal ramifications *(bronchial tubes)* enter into the pulmonary vesicles, which vesicles we have compared to the pores of a sponge.

3. How are the drawing-in of the air (inspiration) and its expulsion (expiration) accomplished?

The lungs fit the thorax inside exactly, and must take part in all its movements. If the thorax is lifted and expanded by the muscles designed for inhaling, the porous tissue of the lungs is also drawn apart, and the air must flow in through the trachea into the lungs. If the chest is compressed by the muscles for exhaling, then the lung is compressed and the air is driven out through the trachea—it is precisely as if one opened a bellows and again compressed it.

4. For what purpose does the air continuously flow into, and out of, the lungs?

In the walls of the fine porous tissue of the lungs run very many minute blood-vessels *(capillaries)*.

The air by which we are surrounded is a mixture of different gases. During inhalation, one of these gases *(oxygen,* or the vital air) penetrates through the fine walls of the blood-vessels into the blood, and mingles with it. In exhaling, other gases, injurious to life, especially *carbonic acid gas,* issue from the blood, and are conducted outwards through the windpipe. To keep itself healthy, and to maintain all parts of the body in a healthy state, the blood must continually absorb fresh air. If the windpipe grow contracted from disease—for instance, from the formation of too much mucus, or from coagulating substances which exude from the mucous membrane (as in croup, diphtheria), or should the fine pores of the lungs be completely filled with blood, or with exudations (for instance, in inflammation of the lungs), then sufficient vital air cannot reach the blood, and when the disease reaches its climax, death from suffocation may occur.

5. How many respirations per minute are usually made by man?

The average, when he is in a quiet condition, is eighteen. In lively movements (running, ascending), in fever, in diseases of the respiratory passages, the number may be more than doubled. In sleep the number may be decreased. Children breathe more quickly, and have a more rapid pulse than adults.

THE ALIMENTARY CANAL, AND ITS GLANDS.

(THE DIGESTIVE SYSTEM.)

1. *What is the Alimentary Canal, and of what parts does it consist?*

The alimentary canal is the canal or tube into which the food is introduced, so that it may pass thence into the blood, and be added to the different parts of the body for their sustenance, and, with children, also for their growth. This canal is divided into five principal sections :

(a.) The mouth with the upper part of the pharynx (the throat). In the mouth the foods are reduced by mastication, and are there mixed and enveloped with mucus and saliva so that they may slide the more easily into the pharynx, and into the œsophagus.

(b.) The œsophagus is a membranous tube lying close in front of the spinal column, behind the larynx and windpipe, and reaches from the throat to the stomach.

(c.) The stomach is the broadest sac-like part of the alimentary canal, lying close below the end of the breast-bone, at the "pit of the stomach." It is capable of very considerable dilatation, but, nevertheless, when empty and healthy, must be able to contract again completely. In the coats of the stomach lie many small glands, which secrete an acid juice *(gastric juice)* very important to digestion ; this juice reduces

the food, and also the coagulated albumen, into a soluble gelatinous condition.

(*d.*) From the stomach the food enters the long *small intestine*, which lies in many convolutions in the abdomen. The *bile*, secreted by the liver and accumulating in a bladder (gall-bladder), flows into the upper part of the small intestine through a narrow channel. Here the excretory duct of the *pancreas* likewise discharges ; the juices so discharged mingle with the chyme coming from the stomach, and are very important to digestion. (If the issue of the bile into the intestine is hindered, it enters the blood, and *jaundice* is the result.) This small intestine is the longest part of the alimentary canal—in an adult of medium size the whole canal is about nine yards in length, between seven and eight of which belong to the small intestine.

(*e.*) At the lower part of the abdomen on the right hand side (region of the right groin) the small intestine discharges itself quite suddenly into the *large intestine*, which is twice its size, and which here has a short sac-like appendix *(cæcum)*, and a finger-long vermicular piece adjoined *(appendix-vermiformis)*. The large intestine (in which the chyme*—by the bile, coloured yellowish-brown—becomes condensed to somewhat solid fæces, because the fluid parts of the chyme [the chyle] were absorbed by degrees by the walls of the intestine) now ascends to the right in the abdomen, under the liver, then runs across to the left, close under the stomach as far as the spleen (above, to the left, in the abdomen), curves thence downwards, runs down as far as the left groin, where it makes some

---

* "*Chyme*—The pulpy mass formed by the food in its first great change in the process of digestion. *Chylification*—The process by which the chyle is separated from the chyme. *Chyle*—The milk-like liquor from which the blood is formed, occupying the lacteal vessels and thoracic duct."—MAYNE.

short convolutions, enters into the pelvis, and runs thence in
the medial line downwards, to the anus. The lower end of
the large intestine, a little longer than the finger, is called the
rectum.

2. *Of what do the walls of the alimentary canal consist?*

They consist of a muscular membrane, and inside, of a mucous membrane, with their own glands—the *intestinal glands.*
The muscular membrane contracts in its length and in its
breadth like an earthworm, and so impels the chyme onwards.
(By the reversed movements of the stomach, of the œsophagus,
and of the throat, vomiting is produced.) The will has no
influence over the greater portion of this muscular membrane. Only at its upper end, the throat, and below, at the
anus, are the muscles the action of which is subservient to
the will.

3. *In what way does the chyle, the juice of food, enter into the
blood?*

A great part of it, especially the water, passes quickly
through the fine walls of the blood-vessels of the intestinal
mucous membrane directly into the blood, and conveys with
it some of the substances already dissolved in the water.
Another part of the alimentary constituents, changed already
by the various digestive juices (saliva, gastric juice, bile, intestinal juice) passes through minute pores into the absorbent
vessels (lymphatic vessels—here in the intestine termed, lacteal
vessels).

These absorbent vessels are spread, net-like, in large
numbers over the interior surface of the intestinal mucous
membrane ; they gradually combine and form larger vessels,
which remain thin and fine, and, in the mesentery, congregate
into larger spongiform bodies *(mesenteric lymphatic glands or
chyliferous glands)* as large as beans ; leave these again, unite,

and finally form a canal as thick as a violin-string, which runs up the vertebral column, and discharges into the great brachial vein above on the left side. Here the chyle flows into, and mingles with, the blood.

4. *How is it that, in their movements, the many convolutions of the long intestinal canal do not become entangled, nor get into the thoracic cavity?*

In rare cases it happens that the intestinal convolutions do so shift and turn under and over each other that the fæces cannot pass further along the intestinal canal, and they are forced backwards and upwards *(fæcal vomiting—miserere).* This cannot easily occur, because the intestine, by a broad ligament *(the mesentery)* fastened to it in every part, is fixed at its back to the front of the vertebral column.

The intestines are prevented from rising into the thoracic cavity because it is divided from the abdominal cavity by a partition—*the diaphragm;* this is attached inside to the lower ribs, to the vertebral column, and to the breast bone, and is extended like a vaulted tent roof. The diaphragm is a muscular membrane; when it contracts, the arch it forms becomes flatter, the thoracic cavity larger, and the abdominal cavity smaller. As the diaphragm contracts during inspiration, it serves materially to enlarge the thoracic cavity and the lungs.

THE URINARY ORGANS.

1. *Of what does the organic system for the secretion and evacuation of urine consist?*

At the back, in the upper part of the abdominal cavity (on the right side, below the liver; on the left, below the spleen) a kidney lies on each side. The kidneys secrete the urine ; this is conducted from them through long thin tubes *(ureters)* lying behind the intestines, which tubes run downwards to the middle of the pelvic cavity into the urinary bladder, which lies below in front in the medial line of the pelvis. From this bladder a canal *(urethra)* runs downwards through which the urine passes from the body. The urinary bladder, like the intestinal canal, consists of two layers—a muscular membrane lined with a mucous membrane. We have as little power over the muscular membrane of the urinary bladder as we have over the muscular membrane of the intestine, but we can open at will the lower opening of the bladder, usually closed (where it passes into the urethra), and, by the abdominal muscles, can exercise pressure upon the full bladder, and so materially assist the outflowing of the urine.

2. *Of what use to the body is the secretion and the evacuation of urine?*

Urine is not pure water, but many other constituents are dissolved therein, which, if they remained in the blood, would become very dangerous to life.

The function of the kidneys is, to draw these injurious constituents from the blood to themselves, and to convey them from the body with the surplus water, as described.

THE POSITION OF THE VISCERA IN THE THORACIC AND ABDOMINAL CAVITIES.

By a muscular-membranous partition, lying transversely, and, inside, fastened to the lower ribs, the vertebral column, and the breast-bone, the thoracic cavity is divided from the abdominal. This partition is termed the *diaphragm ;* it has openings at the back by the vertebral column, through

X

which the great artery and the œsophagus pass from the thoracic into the abdominal cavity, and the great vein passes out of the abdominal into the thoracic cavity.

In the centre of the *thoracic cavity*, at its upper part, lying at the base of the trachea, are the great arteries and veins, which come out of, and return into, the heart. The heart itself is turned somewhat to the left, and is enclosed in a membranous sac *(pericardium)*. This sac is attached underneath to the diaphragm, in front to the sternum, at the back to the vertebral column, and reaches upwards as high as the great arteries and great veins. Thus a central section of the thoracic cavity is formed, leaving a cavity on each side of it,— a right and a left; in each of these a lung is contained.

Abdominal cavity.—If the abdominal wall *(peritoneum and abdominal muscles)* be taken away in front, the intestines are seen, covered mostly by the *omentum.** This is a very fine, net-like membrane, often provided with much adipose tissue, and, at its upper part, is broadly fastened to the transverse-lying portion of the large intestine and to the stomach, and, like an apron rounded-off at its lower edges, hangs in front of the intestines as far as into the lower part of the abdominal cavity. If the omentum be turned upwards, then the convolutions of the small intestines are seen; if drawn somewhat downwards, then, in the centre, the transverse portion of the large intestine becomes visible, and above it the stomach, on the right side of which (partly covering it) is the liver, and at the left side the spleen. Should it now be desired to remove the intestine from the abdominal cavity, this is quickly seen to be impossible until something further has been done, because the intestine is everywhere fastened at the back to the vertebral column by a broad ligament (the

* "The duplicature of the peritoneum."—MAYNE.

mesentery), often very rich in adipose tissue. If the mesentery be cut away from the spinal column, and the intestines be taken out, then above, on each side at the back are the kidneys, and these are covered, in addition, with a white membrane, the *peritoneum*. At its upper part, the peritoneum is attached to the diaphragm ; at the middle, to the mesentery ; underneath, to the pelvis ; and at the front, to the inner surface of the abdominal walls. If it be severed from the kidneys, then the ureters, which pass from them, are seen on each side ; and, in the middle, on the vertebral column, the great abdominal artery and abdominal vein *(vena cava)*. Quite below, in front, in the medial line of the pelvis, is the urinary bladder, and the rectum behind it.

THE HUMAN BODY AS A LIVING ORGANISM.

The life of the human body is dependent, not upon a single organ, nor upon a single system of organs, but upon the united action of all its parts. As, in an ingeniously constructed machine of many parts, no single part is unnecessary, so also in the human body, no part is unnecessary, although one part may be more important than another. And, as a steam engine, which may be described as an inanimate organism, not only stops when not stoked, but also, when the boiler is not tightly closed, or the wheels not oiled, or when single parts are worn out or are rusty, so there are also many causes why a "living organism" like the human body may suddenly, or slowly, come to a full stop.

If we seek for and prepare our food, movement of the arms and legs is requisite ; the action of swallowing is necessary, in order to absorb the food ; the movement of stomach and intestine, in order to carry forward in the body the food absorbed, and to excrete that which is useless and injurious.

The action of the heart is requisite to convey blood to all parts of the body; and the act of respiration in order continually to absorb fresh vital air into the blood. We also require arms and legs to protect ourselves from enemies, to build houses for habitation so that we may be sheltered from the inclemency of the weather, and for other purposes.

What causes the contraction of the muscles? The motor-nerve-power proceeding from the brain. But this can only be continuously renewed when the brain continually receives healthy (refreshed by respiration, renewed by food) blood. The muscles also retain their vital power of contraction only so long as they are perfused by healthy blood.

It would be impossible for the body to exist without the senses; could we neither see, nor feel, nor hear, nor taste, nor smell, we should know nothing of all external to us, and nothing of ourselves. We could not seek nor prepare food for our own use; we could not escape from dangers that might threaten us from our fellows, from animals, from cold, from fire, &c. But the organs of sense could not exist unless healthy blood were continually supplied to them by the action of the heart.

How much is requisite to preserve the blood in a continuously healthy condition so that it constantly re-invigorates the organs of movement, and the organs of the senses! Above all, the inspiration of oxygen, and the supply of food! That the juices from the foods introduced may be mixed with the blood, not only must the minute passages from the intestine to the interior of the blood-vessels be open, but, for their absorption into the blood, the foods must be prepared and changed by means of the various digestive processes.

However, many things besides get into the blood, which, permanently remaining, might become injurious; or again, many used up, or unsuitable substances come from the different parts of the body back again into the blood: these must be expelled from the blood and the body by respiration, by the action of the kidneys, and by perspiration.

Finally, every minute particle of the body which lives and grows must have the power to attract certain substances from the blood to itself, and to make them that which it itself is. From the blood conveyed to the muscle it must again form muscular tissue if it is to exist, or to increase; the salivary glands must produce, not only saliva from the blood sent to them, but must also take from it the substances of which they themselves consist; the bones must attract to themselves, from the blood, those constituents by which they are maintained and are enabled to grow, and so on.

The stone increases only when new stone masses are deposited upon it; it can become larger, but it does not grow of itself—it does not live. Such bodies only are described as *living* which possess the capacity, from their own inherent power, to absorb dissimilar constituents, and to develop them into that which they themselves are. The steam engine is an inanimate organism—it performs enormous work when set in motion by means of heat, but the fuel never becomes a component part of the engine—the coal never becomes iron. By means of its fuel—food—the living organism is not only set and maintained in motion, but it grows and develops ever anew, because it transforms the constituents of its food into its own constituents.

Wonderful as is the construction of the living organism, the body, yet it is only capable of working for a certain length of time—this may extend to a hundred years, and somewhat over. The derangements in its functions are, at least, just as manifold as are the parts themselves of which the body consists.

What is termed *the life of the body* is, therefore, the result of the combined action of many very complex processes. If we are able to give a few glances into these processes, yet we are still far removed from knowing them fully, or from being able to comprehend their interdependence.

FINIS.

INDEX.

is as readable as a novel. For clearness of teaching, for comprehensiveness of the whole field of sickness, and the common-sense principles laid down by Dr Billroth, there is no book in the English language that can be compared to this, and this book should be found in every Household, and in the hands of every Professional Nurse."

The Daily News.

"Dr Billroth's book, based as it is upon his long practical experience in connection with important offices, is likely to prove of greater practical use than most manuals or compilations on the same subject."

The Western Daily Mercury.

"At a time when the demand for properly qualified and thoroughly trained nurses is growing so rapidly, the appearance of Dr Billroth's Handbook is welcome. Dr Billroth is a man of considerable eminence in his profession in the Austrian capital, being President of the Imperial and Royal Medical Association, and Director of the Surgical Institute of the Vienna University, and therefore speaks with authority. The average reader will not fail to recognize the great practical help which a careful study of the volume must afford to any lady who is contemplating the adoption of Nursing as a profession, or has already entered upon it ; while the mother of a family who should be fortunate enough to consult its pages will certainly find her labour repaid in an intelligent equipment for the emergencies of the sick-room. Dr Billroth enters, in great detail and in the most systematic manner, into the various branches of a nurse's work, whether in the home or in the hospital. His mode of treating these subjects is technical where necessary ; but he continues at all points to keep his language within the easy comprehension of the average layman, and his Translator does him everywhere full justice."

The Canterbury Press.

"Liability to sickness is the common lot of humanity. Sooner or later, for brief or for long periods, it must be endured ; hence, how to secure the most intelligent Nursing has become one of the leading questions of the day. In many thousands of cases the attendants upon the sick are at a loss to know how best to nurse them. To supply this knowledge we now have a book pre-eminently fitted for use by everybody, one which it is impossible to read without the reader realizing what is best to be done in sudden illness, in temporary or in chronic diseases, and how to do it. Dr Billroth is one of the first surgeons in Europe, and his experience with disease in every form is as large as that of any medical man in Europe. In this book he has embodied the results of his public and private practice in such a way that we know of no other book on the subject that can be compared with it for the lucidity of its details and for its comprehensive grasp of illness in its varied forms. We congratulate the Translator upon the very satisfactory way in which this part of the work is fulfilled. Wherever there is illness this book is invaluable—it is what it professes to be, 'A Handbook for Families and for Nurses,' and should find a place in every household."

"This is a book which will do no harm in the hands of the veriest tyro, while in the hands of the skilled nurse, or of the intelligent student of the labours of 'The Beloved Physician,' it should prove invaluable. Dr Billroth, the embodiment of the latest discoveries in the healing art, has done well to promulgate an authorized translation of his book among us."

"Many books on Nursing have appeared of late years, some good, others the reverse; but by far the best book we have lately come across is that entitled 'The Care of the Sick,' by Professor Billroth, of Vienna. No detail of Nursing that could conduce to the patient's welfare or comfort has been omitted—whether for Professional Nurses, or for the Heads of Families who have to nurse the sick in their own homes, Professor Billroth's book contains all the needful information that the most exacting could require. It is very finely Illustrated."

"This work is well known on the Continent, and has already gone through three editions. Dr Billroth, as Professor of Surgery in Vienna, had a wide experience of the different methods employed in nursing the sick, and, in consequence, 'realised,' as Miss Endean tells us, 'how little, comparatively, is known of the essential requirements of the sick-room; he, therefore, wrote this book for the instruction and guidance of all interested in the care of the sick, stating therein with much clearness, the causes, nature, and symptoms of various diseases, and the main principles of good sick-nursing.' From these instructions a mother will learn not only what is necessary to the care of a sick member of her family, but also the best means to adopt for the prevention of sickness, and for the maintenance of health in her household. But this volume is not written simply for mothers and other members of a household who have inevitably, on occasions, to play the nurse's part. For those—and they are increasing with great rapidity—who desire to train as professional sick-nurses, 'the contents of this Handbook should,' in the opinion of its author, 'constitute the First Lectures (of the Preparatory Course) on the Care of the Sick.' From the practical point of view, Dr Billroth's work is admirable, being lucid and concise in the instruction it gives under a variety of titles, such as 'The Sick-Room,' 'General Rules for the Care of Patients Confined to their Beds,' 'Preparations for Operations and Bandaging,' 'Nursing in Epidemics,' 'Care of Nervous Patients,' 'Aid in Accidents,' and 'Food and Diet.' Altogether, this is a valuable book, and Miss Endean's translation has been admirably executed."

Sampson Low, Marston & Co.'s Publications

The Nursing Record

A JOURNAL FOR NURSES.

And a Chronicle of Hospital and Institution News, &c., &c.

ONE PENNY.   **Every Thursday.**   **ONE PENNY.**

TWENTY-FOUR PAGES.

COMMUNICATIONS from all parts of the country are cordially invited, and liberal arrangements are made for reprints of original articles, and for such illustrations as serve to increase their value or interest. Reports of Nursing Vacancies, Appointments, Meetings of Societies, &c., and Newspapers, &c., containing (marked) accounts of matters of local or personal interest or importance, will be gladly received. Correspondence upon all subjects associated with Nursing specially invited.

TERMS OF SUBSCRIPTION.

The **NURSING RECORD** can be had by sending Postal Order or Stamps to Messrs Sampson Low, Marston & Co., Limited, the Proprietors, St Dunstan's House, Fetter Lane, London, to whom all money payments should be made.

For One Year, post free, to any part of Great Britain and Ireland,	.	. .	6s. 6d.
Six Months,	do.,	do.,	3s. 6d.
Three Months,	do.,	do.,	1s. 9d.
To America and the Continent, the Annual Subscription, including postage,	.	. . .	10s.

LONDON:
SAMPSON LOW, MARSTON, & Co., Limited,
St Dunstan's House, Fetter Lane, Fleet Street, E.C.

Fifth Edition, at all Booksellers and Libraries.

In Darkest Africa. Being the Official Publication recording the QUEST, RESCUE, and RETREAT of EMIN, GOVERNOR of EQUATORIA. By HENRY M. STANLEY, LL.D., etc., Author of "How I found Livingstone," "Through the Dark Continent." With numerous Original Illustrations and many Maps. 2 vols. demy 8vo, cloth, 42s.

"The style of the narrative is direct, vigorous, and incisive, as beseems one who is a man of action rather than a man of letters; but in many of the descriptive parts Mr Stanley shows that, consummate man of action as he is, he is a born man of letters as well."—*Times.*

Supplementary Volume to Mr Stanley's " In Darkest Africa."

Emin Pasha and the Rebellion at the Equator. A Story of Nine Months' Experiences in the last of the Soudan Provinces. By A. J. MOUNTENEY JEPHSON, one of Stanley's Officers. Written with the revision and co-operation of HENRY M. STANLEY, D.C.L., etc. A Preface also by Mr STANLEY. With Map and numerous Illustrations. Third Edition. Demy 8vo, cloth extra, One Guinea.

How I found Livingstone; including Four Months' Residence with Dr Livingstone. By HENRY M. STANLEY, D.C.L., &c. With Maps and Illustrations. Crown 8vo, cloth, 3s. 6d. The Original Edition, superior in paper and binding, price 7s. 6d., can still be obtained.

Through the Dark Continent, from the Indian to the Atlantic Ocean. By HENRY M. STANLEY, D.C.L., &c. With Maps and Illustrations. Crown 8vo, cloth, 3s. 6d. The Original Edition, superior in paper and binding, and with the original Maps, 12s. 6d., can still be obtained.

Through Masai Land: Exploration among Snowclad Volcanic Mountains and Strange Tribes. By JOSEPH THOMSON, Author of "To the Central African Lakes and Back," &c. New Edition, with numerous Illustrations. Crown 8vo, cloth, 7s. 6d.

The Heart of Africa: being Three Years' Travels and Adventures in Unexplored Regions. By GEORGE SCHWEINFURTH. New and Cheaper Edition, with Map and Illustrations. 2 vols. crown 8vo, cloth, 3s. 6d. each.

Two Kings of Uganda; or, Life by the Shore of the Victoria Nyanza. By ROBERT P. ASHE, M.A., late of the Church Missionary Society's Nyanza Mission. With Illustrations and a New Map of Eastern Equatorial Africa. New Edition, with information relating to Uganda brought down to date. Crown 8vo, cloth, 3s. 6d.

LONDON:

SAMPSON LOW, MARSTON, & Co., Limited,

St Dunstan's House, Fetter Lane, Fleet Street, E.C.

www.ingramcontent.com/pod-product-compliance
Lightning Source LLC
Chambersburg PA
CBHW021117270326
41929CB00009B/924